# Edwardian Steam

## A LOCOMOTIVE KALEIDOSCOPE

Sectional general arrangement drawing (elevation & plan) for the first (1906) batch of North British Railway 4-4-2s, which ranked amongst the most impressive locomotives of the Edwardian period. Heralding the appearance of a new locomotive type, drawings like this were frequently published in such periodicals as *The Engineer*, *Engineering*, *The Railway Engineer*, and occasionally in *The Railway Gazette* and *The Locomotive Magazine*. *Author's collection*

SCALE—1, INCH TO THE FOOT.

# Edwardian Steam

## A LOCOMOTIVE KALEIDOSCOPE

Philip Atkins

crecy.co.uk

First published 2020

© Philip Atkins 2020

ISBN 9781910809655

Printed in Bulgaria by Multiprint

**Crécy Publishing Ltd**
1a Ringway Trading Estate, Shadowmoss Rd, Manchester M22 5LH
**www.crecy.co.uk**

*Front cover:* The unique Great Western Railway 4-6-2 No. 111, *The Great Bear* (1908) at Paddington Station, c. 1910. *R. Brookman/Rail Archive Stephenson*

*Rear cover:*
*Inset top:* London & North Western Railway Precursor class 4-4-0 No. 737 *Viscount* awaits departure from Euston. *Former Ian Allan Archive*
*Inset middle:* Superheated North Eastern Railway three-cylinder Class Z 4-4-2 No. 729 stands on the King Edward Bridge, spanning the River Tyne between Newcastle and Gateshead. Probably taken when the engine was brand new, with original brass-capped chimney, in September 1911. *R. J. Purves/Rail Archive Stephenson*
*Inset bottom:* Great Eastern Railway non-superheated 0-6-2T No. 1000, constructed at Stratford Works during 1914. *Former Ian Allan Archive*

*Rear cover main:* The prototype for a class that totalled 130 locomotives, London & North Western Railway four-cylinder 4-6-0 No. 2222 *Sir Gilbert Claughton* (with indicator shelter attached), is inspected by the LNWR Locomotive Committee at Crewe Works on 7 March 1913. *Science and Society Picture Library*

**Picture credits**
Every effort has been made to identify and correctly attribute photographic credits. Should any error have occurred this is entirely unintentional.

# Contents

# 1

# A Golden Age?

The Edwardian era, by convention usually defined as 1901–14, i.e. including the four years beyond the death of King Edward VII, has only fairly recently passed beyond human recall. It is now commonly perceived as having been a Golden Age, which socially speaking it certainly must have been for a highly privileged but relative few. It undoubtedly also was from the contemporary railway viewpoint, simply on account of its now unbelievable individuality and diversity. This arose from the simultaneous existence of approximately 120 legally independent companies, which operated both in intense competition against and also in close cooperation with each other. Their full extent and inter-relationships, actually shown as for the year 1921, is graphically portrayed in the admirable *British Railways Pre-Grouping Atlas and Gazetteer*, that was first published by Ian Allan Ltd in 1958, and which has been regularly reprinted since.

These numerous enterprises were dominated by fourteen major companies, eleven English and three Scottish, so rated according to the simple criterion of each having a total locomotive stock that exceeded five hundred locomotives. In 1913 these fourteen companies collectively accounted for 90 per cent both of the total route mileage and also of the national main line locomotive stock. In fact, the three largest of these, the Great Western, the London & North Western and the Midland, between them operated 30 per cent of the total route mileage, and nearly 40 per cent of the locomotive stock, which individually on each had recently topped the three thousand mark. That said, the total route mileage of the seemingly far-flung Midland, which despite the fact that it stretched from London to Carlisle, and also boasted a detached presence in South Wales, was in reality surprisingly somewhat less than that of the more parochial North Eastern, which essential extended between the rivers Humber and Tweed, and to the east of the Pennines.

What wouldn't one now give for the services of a time machine, and where better to set it up, wound back say to about 1912, than Carlisle Citadel Station, where an unparalleled complexity existed in the immediate surrounding area that even London could not rival? At Carlisle terminated not only three of the five largest English, but also all three of the three largest Scottish railway companies, which were all joined by the small, local Maryport & Carlisle Railway. A close second choice would surely be York. Although dominated by the North Eastern Railway, whose administrative headquarters were located in the city, some locomotives of not only the NER itself, but also of the Great Northern and Lancashire & Yorkshire railways

Carlisle, Railway Clearing House map, 1912.

were stationed there, as were also a handful from the Great Eastern, and at times even the odd one from the Great Central, while Midland locomotives would also put in appearances working from Leeds. In its extent the national network was approaching its peak by 1900, following the recent completion of the final main line, the Great Central Railway's London Extension. In reality, very much in the spirit of the late 19th century, the latter route had originally been envisaged by Sir Edward Watkin merely as but a part of his grandiose, although ultimately unrealised, project to link Manchester directly with Paris by rail, also via the South Eastern Railway and by means of a tunnel passing under the English Channel.

The most evocative and detailed published first-hand accounts of this period concerned what were arguably the two least likely of the fourteen major railways. The progressive Lancashire & Yorkshire Railway was one of only two major English railways that did not terminate in London, while in Scotland the smaller and rather unassuming Glasgow & South Western Railway existed very much in the shadow of the larger and more flamboyant Caledonian Railway. But for the respective diligence of Eric Mason and David L. Smith our present-day knowledge of LYR and GSWR locomotives and their workings would be extremely limited.[1, 2]

By 1913, when the total national steam locomotive stock was *precisely* quantified for the first time, this was by then only a few hundred short of its ultimate all-time peak of just above 24,000, which would be operative ten years later between 1921 and 1924. At no time did the British steam locomotive develop more rapidly, and more diversely, than during the Edwardian period. To take a particularly extreme example, little more than six years elapsed between the completion in October 1901 by the Great Northern Railway of the final British 'single-driver' 4-2-2, a type that by then was already effectively obsolescent, and the emergence in February 1908 of the first British 4-6-2 or 'Pacific' on the Great Western Railway, which was distinctly ahead of its time as far as Britain was concerned. Indeed, neither this engine nor anything like it was ever replicated by the GWR. A more practical example was the self-styled 'Premier Line', the London & North Western Railway, on which in 1903 the latest express passenger locomotive class was the 57½-ton, non-superheated, four-cylinder compound Alfred the Great 4-4-0. Only ten years later, in 1913, appeared the new 77¾-ton superheated four-cylinder simple Claughton 4-6-0. This fascinating period produced more than 250 new locomotive designs. In addition to these several more were authorised, or at the very least initiated, before 1915, which appeared thereafter and so have also been considered to be 'Edwardian' as to their origin.

Great Northern Railway 4-2-2 No. 265, built in October 1901, heads a train formed of very mixed passenger rolling stock. *Former Ian Allan Archive*

[1] Mason, E. *The Lancashire & Yorkshire Railway in the Twentieth Century*, Ian Allan Ltd, 1954.
[2] Smith, D. L. *Locomotives of the Glasgow & South Western Railway*, David & Charles, 1976.

GWR 4-6-2 No. 111, *The Great Bear*, built in February 1908, seen in its later years with plain cast iron chimney, works a heavy Cheltenham express west of London. *The Great Western Trust*

This kaleidoscopic array was enriched by the appearance of several entirely new locomotive wheel arrangements, both tender and particularly tank, which at their most extreme included the solitary British 0-10-0 examples of each. It also included the 'top secret' scarcely ever photographed Paget 2-6-2 tender engine No. 2299 on the Midland Railway. There were several experiments on four different railways with compounding on express passenger 4-4-2s during 1903–06, some of which having been built do not then seem to have been evaluated with any noticeable zeal. This veritable cavalcade was effectively 'bookended' by the debut of the Victorian era-designed ornate South Eastern & Chatham Railway Wainwright Class D 4-4-0 in February 1901, and at the other extreme by the eventual completion in December 1919, after 5½ years, of the unique Midland Railway four-cylinder 0-10-0 banking engine No. 2290, whose construction had begun back in May 1914. Interestingly, the talented locomotive designer James Clayton had a close involvement in each of these two extremes, *and* he designed the Paget locomotive almost single-handed.

Very early on, 1903 was a quite exceptional year in terms of the number of new locomotive designs that appeared, amounting to no fewer than thirty in total. Most notable among these included the ground-breaking prototype American-style Churchward taper boiler 4-6-0 and 2-8-0 tender, and 2-6-2 tank locomotives on the Great Western Railway. Some of their instantly recognisable descendants were still very much in evidence on British Railways' Western Region sixty years later. Express passenger 4-6-0s also made their first appearances on the Caledonian, Glasgow & South Western and Great Central railways, and 4-4-2s on the North Eastern and (again) the Great Central. Certain other new arrivals that year were rather less memorable, not least the rakish compound 4-6-0 swansong of Francis Webb on the London & North Western, and the cumbersome Henry Hoy inside-cylinder 2-6-2 tank engines on the Lancashire & Yorkshire, which were both doomed to unusually early extinction. The year 1906 was a notable one for particularly diverse new express passenger locomotives, while 1913 was also a memorably productive year. By this time the British steam locomotive, in addition to its very recent enhancement by the increasingly rapid adoption of the superheater, was also perceptibly just beginning to assume its ultimate format. This would later be enshrined in the short-lived British Railways Standard steam locomotives built forty years later in the 1950s. The keynote was two outside cylinders in association with external Walschaerts valve gear, all set beneath a high running board in order to maximise accessibility for routine servicing purposes on shed.

Aesthetically speaking, Edwardian locomotive designs were almost invariably handsome to behold, with their form very often transcending their function, while their performances did not always live up to their fine appearances. They were designed in detail by largely anonymous draughtsmen, who had not always necessarily experienced the sharp end of actual locomotive *operation*, particularly if they had been trained in a commercial locomotive works, rather than on a working railway itself. On the other hand, during the period under review here, many of those working in branches of the commercial locomotive-building industry, which was then at its zenith, and particularly those located in Glasgow, Manchester and Darlington, would have been more familiar than their neighbouring purely railway-based counterparts with such 'advanced' features as Walschaerts valve gear and large boilers having wide fireboxes. Such characterised some of the locomotives that they were already designing for service overseas, notably in India, South Africa and South America, well before they became more familiar in Britain after say, 1918.

Some of these draughtsman would have been involved in the design of the new standardised locomotives for India, inaugurated in 1902, whose precise paternity appears never to have been fully identified. Some early examples of these inside-cylinder 4-4-0s and 0-6-0s, and outside-cylinder 4-4-2s, 4-6-0s and 2-8-0s, would not have looked out of place back home, on the Great Central Railway in particular. Something of a mystery attaches to the underlying reasons for something of a three-cornered resemblance between the Glasgow & South Western Railway 4-6-0s, the Indian 'Mail' 4-6-0s, and certain 4-6-0s on the Great Central.[3]

---

[3] Atkins, P. Private locomotive building and the Indian connection, in *Bedside Back Track*, Atlantic Transport Publishers, 1994, pp.19–24.

One of the earliest applications of outside Walschaerts valve gear to a British main line locomotive, as fitted to Great Northern Railway four-cylinder 4-4-2 No. 271 in April 1904. Fifty years later Walschaerts valve gear would prominently feature on about 30 per cent of the 18,500 steam locomotives then in operation on British Railways in 1954.
*Science and Society Picture Library*

There was, in the furtherance of their modestly paid careers, a measure of interchange of drawing office personnel, both very senior and quite junior, between the commercial locomotive builders and the various large railway works. Pre-1903, Neilson & Co. in Glasgow was often regarded as the Oxbridge of locomotive engineering. As of 1910, the respective locomotive chief draughtsmen of the North Eastern and South Eastern & Chatham railways, George Heppell and Robert Surtees, who evidently were very close friends, were both ex-Robert Stephenson & Co. from the latter's pre-1900 years when still based in Newcastle. Their opposite number on the Great Central, William Rowland, had previously worked for both Beyer, Peacock & Co. and the Vulcan Foundry, not to mention the Lancashire & Yorkshire Railway at Horwich. Between 1906 and 1914 all three simultaneously oversaw the detailed design of numerous extremely imposing, albeit often coal-hungry, locomotives that were very much of their time. In 1913, the London & South Western Railway recruited the chief designer at the North British Locomotive Company, Thomas Finlayson, as its chief locomotive draughtsman, an appointment that heralded a radical change in locomotive design at Eastleigh after the sudden death of Dugald Drummond. Conversely, in the early 20th century Beyer, Peacock & Co. regularly appointed former locomotive superintendents recruited from British and Irish railways when seeking a new managing director, i.e. Henry Hoy (Lancashire & Yorkshire Railway), Edward Watson (Great Southern & Western Railway), and Robert Whitelegg (latterly Glasgow & South Western Railway).

Many British and Irish locomotive superintendents were members of the Association of Railway Locomotive Engineers (ARLE), which had been formed in 1890. This body traditionally met twice a year, in November in London at the Midland Grand Hotel, St Pancras, and elsewhere up and down the country in June. For example, in June 1908 the venue was Harrogate, in 1909 Windermere, in 1910 Chester, in 1911 Glasgow, in 1912 Aberdeen, in 1913 Tunbridge Wells, and in 1914 Barrow in Furness.[4] Typically at any one time the total membership numbered around thirty-six, of whom an average of twenty members attended the winter meetings in London during this period, compared to only fourteen at those meetings that were held elsewhere during the summer. The 'country' meetings were sometimes chaired by a member with a fairly local association, for example Matthew Stirling (Hull & Barnsley Railway) who presided in Harrogate in 1908. Interestingly, Stirling had the unparalleled and remarkable distinction of serving a single railway company, the HBR, as its locomotive superintendent for no fewer than thirty-seven years, from 1885 until 1922.

[4] G. Hughes to author, March 1993.

*Above and opposite:* Only ten years separated the construction by the South Eastern & Chatham Railway at Ashford Works of Class E 4-4-0 No. 19 in 1907, seen in the full glory of its elaborate green livery (above), and the prototype Class N 2-6-0 No. 810 in 1917 in wartime austerity grey paint (opposite). The latter had actually been designed, at least in outline, three years earlier, in mid-1914. *Former Ian Allan Archive*

He had also attended the inaugural meeting of the ARLE at St Pancras in January 1890, together with his father Patrick (GNR) and uncle James (South Eastern Railway), and is on record as having attended, at the very least, every one of the twenty-three meetings that were held between November 1907 and November 1918. Another enthusiastic member from one of the smaller railway companies was William Pettigrew of the Furness Railway, who hosted the meeting at the Furness Abbey Hotel, which was owned by the Furness Railway, at Barrow in June 1914.

At such meetings technical issues of mutual interest would be discussed, for example harmonising the method of calculation of superheater element heating surface, i.e. whether this should be internal or external and be confined to only that portion within the flue tubes, or what percentage of the boiler working pressure should be taken when calculating locomotive starting tractive effort, and should a deduction be made for the piston rods? The possibility of standardising tyre profiles was another major topic. Most of the delegates would therefore have been personally reasonably well acquainted with each other at any given period. In 1912 the secretary of the ARLE was Harry Wainwright of the SECR, who also chaired the June 1913 meeting in Tunbridge Wells, only very shortly before his effective dismissal at Ashford by the SECR directors. A little later, during 1917–18, serious but ultimately abortive attempts were made to formulate a range of proposed national standard locomotive designs, for construction railway wide after the war had ended.[5] These particular deliberations were chaired by George Churchward from the GWR, who during the years immediately before the war had scarcely attended any ARLE meetings at all. Nor were these early 20th-century meetings seemingly graced by the presence of either of the two Drummond brothers, Dugald and Peter.

A further body, that came to embrace both the most senior and most junior members of the British locomotive engineering profession, both commercially and railway based, was the Institution of Locomotive Engineers, which came into being in 1911. It began as an offshoot of the Stephenson Locomotive Society, which itself had only been founded in 1909. The Institution held regular meetings in different parts of the country, and even abroad, especially in South America, at which technical papers were delivered, later to be published in its bi-monthly journal. The latter gave an unparalleled insight into contemporary, although essentially British, locomotive and railway rolling stock design philosophy. The Institution would cease to exist as an independent body in 1969, shortly after the demise of steam working on the British national railway network.

In an age when both coal and labour were still relatively cheap in real terms, British locomotives were almost invariably finished in very attractive, sometimes even elaborate, liveries. Generally speaking they were kept extremely clean, as is evidenced by the numerous

[5] Atkins, P., Much ado about nothing (British national locomotive standardisation proposals, 1917–18), *Back Track*, August 2008, pp. 461–7.

contemporary photographs that have happily survived from that period. But there were already signs on the South Eastern & Chatham Railway, for example, that the glories of the Edwardian era were beginning to slip away, even before the events of August 1914. Latterly, the hitherto highly polished brass domes of the elegant Wainwright 4-4-0s were beginning to be painted over, and in the interests of increased route availability, their tall copper-capped chimneys were being replaced by slightly shorter plain iron castings. Also, by 1914 the elaborate panelling of the luxurious Brunswick green livery was sacrificed in favour of single yellow lines, while the SECR armorial device formerly emblazoned on the driving wheel splashers was now becoming omitted.

Finally, during the weeks before the fatal shot was fired in Sarajevo that unwittingly triggered the abrupt end of the Edwardian era, a modern outside-cylinder 2-6-4 passenger tank engine, and a 2-6-0 mixed traffic/goods engine directly derived from it, were outlined by the SECR at Ashford. In their general appearance, with external Walschaerts valve gear and taper boilers, these remarkably anticipated the British Railways Standard steam locomotives of the 1950s. On account of the war, completion of the two prototypes was considerably delayed until mid-1917, when these emerged in plain grey livery. The 2-6-0s were happily destined to endure almost to the end of steam working in southern England nearly fifty years later. For their part, the 2-6-4 tanks as such, later to be named by the Southern Railway after local rivers, would enjoy an overall currency of only ten years. Indeed the final examples operated as built for only a matter of months. All were rebuilt as now anonymous 2-6-0 tender engines in the wake of the fatal derailment at Sevenoaks in August 1927 in which one of them had been involved. This event cast a shadow, by no means for the first time, on the employment of large tank locomotives on fast passenger duties.

## Contemporary Edwardian railway and locomotive literature

In more recent years the British main line steam locomotive has been retrospectively chronicled in very considerable detail, particularly under the auspices of the Railway Correspondence & Travel Society, yielding information that is far beyond what would have been available to any contemporary Edwardian observer. Then, he (or she) would have been largely reliant upon the monthly *The Locomotive Magazine*[6] and *The Railway Magazine*, which began publication almost simultaneously in 1896 and 1897 respectively, utilising the newly developed printing technology whereby photographs, or 'half tones', could be reproduced amid printed text. Their appearance might also have reflected an increase in public interest in railway matters following the second so-called 'Railway Races to the North' in 1895. For the professional, in-depth technical details of the latest locomotives, including sectional general arrangement drawings, were regularly featured in *The Railway Engineer* (published from 1880), and the two well-established general engineering 'heavies', *The Engineer* (1856–) and *Engineering* (1865–). The weekly *Railway Gazette* (1905–) contained a mix of up to the minute commercial and technical information. Today, back runs of each of the above serial publications can be consulted via Search Engine at the National Railway Museum in York.

[6] During 1896 *The Locomotive Magazine* was published under the title of *Moore's Monthly Magazine*.

# 2
# Putting the Edwardian Railway Era into Some Perspective

The British railway network began to develop on a purely piecemeal basis during the 1830s. From the beginning of the 1860s until the eve of the First World War its progress was statistically chronicled in considerable detail, railway company by railway company, large and small, and summarised by grand totals, in the *Board of Trade Annual Railway Returns*. Extracted from these volumes and given below are the most fundamental *combined* statistics at the beginning of the six decades.

Railway operations in Britain were significantly curtailed in early 1912 by the thirty-seven-day miners' strike that lasted from the end of January until early March. For 1913, following long debate, the format of the annual statistics was extensively revised. As a result this did not always make for strictly valid direct comparisons with those from the previous fifty odd years. For instance, the *total* locomotive stock of individual railway companies was now reported, whereas many of

## Table 1 British railway assets, 1861–1911

|  | 1861 | 1871 | 1881 | 1891 | 1901 | 1911 |
|---|---|---|---|---|---|---|
| Total route mileage | 9,447 | 13,338 | 15,737 | 17,328 | 18,870 | 19,874 |
| **Reported totals:** | | | | | | |
| Locomotive stock | 5,804 | 10,008 | 13,121 | 16,155 | 20,890 | 21,849 |
| Carriages | 13,755 | 21,182 | 29,102 | 37,497 | 46,934 | 48,471 |
| Wagons * | 176,878 | 267,137 | 377,483 | 537,675 | 679,096 | 724,273 |

* This total relates to railway-owned wagons only, and therefore does not include private owner wagons, of which there were no fewer than 700,000 registered by 1925.

## Table 2 British railway performance, 1861–1911

|  | 1861 | 1871 | 1881 | 1891 | 1901 | 1911 |
|---|---|---|---|---|---|---|
| **Passenger:** | | | | | | |
| Train mileage | 49.6m | 83.8m | 119.1m | 163.1m | 213.7m | 258.5m |
| No. passengers | 163.0m | 359.7m | 605.0m | 822.3m | 1,145.5m | 1,295.5m |
| Passenger revenue | £12.4m | £19.4m | £28.6m | £33.2m | £44.6m | £46.3m |
| **Goods:** | | | | | | |
| Train mileage | 49.5m | 86.6m | 115.9m | 144.9m | 167.7m | 151.4m |
| Merchandise (tons) | 29.2m | 64.7m | 68.1m | 85.6m | 114.4m | 108.5m |
| Minerals (tons) | 63.4m | 101.8m | 173.3m | 220.3m | 296.5m | 462.4m |
| Total tonnage | 92.6m | 166.5m | 241.4m | 305.9m | 410.9m | 570.9m |
| Goods revenue | £14.7m | £25.5m | £35.3m | £41.8m | £51.3m | £61.2m |
| **Passenger + goods** | | | | | | |
| Total train mileage | 99.1m | 170.4m | 235.0m | 308.0m | 381.4m | 409.9m |
| Gross revenue | £27.1m | £44.9m | £63.9m | £75.0m | £95.9m | £107.5m |

Heading an East Coast express, Great Northern Railway large-boilered 4-4-2 No. 273, built at Doncaster Works in 1905, picks up water at speed from Scrooby water troughs, in North Nottinghamshire, which had been laid c. 1901. Pioneered on the LNWR in 1861, the extensive provision of water troughs on several railways after 1900, although not in Scotland, greatly enhanced and permitted the acceleration of express passenger services. By 1905, albeit under exceptional circumstances, this simple facility had even permitted *non-stop* running of almost 300 miles between London and Carlisle, and *through* runs approaching 400 miles by a single locomotive between Manchester and Plymouth. *Former Ian Allan Archive*

these had hitherto only reported their *capital* stock, having not acknowledged their *duplicate* stock. This explains why the national grand total appeared to leap from 21,969 at the end of 1912 to 23,664 as of 31 December 1913, to give a seeming increase of 1,695. This was despite the fact that an estimated total of only 433 new engines had been built during 1913, during which approximately one hundred fewer withdrawals occurred, thereby giving an actual net increase of merely about one hundred engines. The true 1912 total had therefore been under-reported by about 7 per cent.

The statistics now also tabulated total *engine* miles in addition to *train* miles, thereby revealing that *overall* direct revenue-earning mileage amounted to only about two-thirds of the total, the balance being accounted for by shunting operations, empty stock, banking and ballast train working, etc. When averaged across the board, total *train* miles amounted to only 69 per cent of total *engine* miles, or roughly two-thirds, but individually ranged from

only 52 per cent on the Lancashire & Yorkshire Railway with its heavy mineral traffic, to 80 per cent on the London Brighton & South Coast Railway, with its predominant passenger and relatively light goods traffic.

Taking Britain's railways as a whole in 1913, 252.2 million passenger train miles generated £54.6m in gross receipts, or yielded an average of 21.6p per mile, while 134.6 million goods train miles earned a corresponding total of £64.0m, or 47.6p per mile. Although goods working accounted for only 35 per cent of the train mileage, it generated 54 per cent of gross traffic revenue. While goods, and particularly mineral traffic, was more remunerative per train mile, this also incurred higher operating costs, due to higher fuel consumption, and more non-remunerative associated shunting and empty train working.

Several of the key statistics given above show a distinct levelling off between 1901 and 1911, when compared with 1891–1901. Whereas during the former period passenger train mileage increased by 21 per cent, peaking around 1911, the resultant revenue increased by merely 5 per cent. On the other hand, despite a significant reduction in goods train mileage since its peak of 175 million in 1900, goods revenue nevertheless increased by 20 per cent, and actual tonnage by no less than 40 per cent. Having said that, general merchandise tonnage surprisingly declined by 4 per cent, yet mineral tonnage (coal, coal products, iron ore and limestone, etc) increased by more than 50 per cent, and would undoubtedly have steadily continued to increase beyond 1913 but for the outbreak of the First World War in 1914. This could be attributed to the continuing annual increase in the mining of British coal up to that point, which would actually peak in 1913 at 287 million tons, when the associated still-burgeoning export coal market amounted to 73 million tons. At that time an eventual annual coal output of 400 million tons was confidently expected in due course!

To handle their substantial share of the export coal traffic, two key players, the Great Central and the Lancashire & Yorkshire railways, which respectively served the rival east coast ports of Immingham and Goole, almost simultaneously proposed unusually large ten-coupled heavy mineral locomotives in order to reduce their operating costs, probably spurred by the recent sharp increase in the price of coal. Regrettably, neither of these giants would ever see the light of day owing to the sudden tragic turn of international events, and the consequent four-year suspension of coal shipments to Europe and elsewhere. After 1918 the export coal traffic would return, and even initially rise above its heady pre-1914 levels, peaking at 79 million tons in 1923 before then declining significantly.

The Board of Trade statistics, having been suspended during 1914–18, were restored in 1919, when the all-embracing concept of 'transport' was the new watchword. Indeed, May 1919 saw the appointment of the first Minister of Transport, Sir Eric Geddes. Geddes had formerly been the Deputy General Manager of the North Eastern Railway, before his secondment to the government in 1916 by the Prime Minister, David Lloyd George. From then on, road traffic competition, which initially utilised numerous military motor vehicles that had suddenly been rendered surplus with the return of peace, began to have an impact on the railways, and under very numerous independent operators later came into its own during the General Strike in May 1926. Immediate post-war plans for new light railways to serve sparsely populated areas of the country, particularly in remote areas in Scotland, were very quickly abandoned. Certain main line railways continued to make modest progress with electrification. However, the ambitious proposal by the North Eastern Railway in 1919 to electrify the East Coast Main Line between York and Newcastle (and even possibly ultimately beyond through to Edinburgh over the North British Railway north of Berwick), failed to materialise owing to the downturn in the regional economy. It had even rather prematurely completed a prototype 4-6-4 electric high-speed passenger locomotive, ominously numbered 13, at Darlington in 1922.

Viewed with the benefit of hindsight more than a century later, the year 1913, together with the dozen years that preceded it, can now be seen to have been the zenith of the British steam-operated railway, at a period when there was no apparent *serious* competition from alternative forms of either motive power or of land transport.

# 3

# Edwardian Locomotive Building

King Edward VII's actual reign, 1901–10, coincided with the peak of steam locomotive construction in Britain. The decade witnessed the completion of an estimated grand total of 15,400 steam locomotives being built by and for the home railways in England, Scotland and Wales, by commercial builders also for export overseas on a variety of rail gauges, and for industrial enterprises both at home and abroad.

Although this *total* figure was some 8 per cent up on that estimated for 1891–1900 (14,250), the number of locomotives built for the *home railways* amounted to only 5,400 as compared to 7,900, a reduction of no less than one third.[1] Following the close of the 19th century, in 1900 total revenue-earning train mileage quite suddenly ceased its hitherto remorseless and almost unbroken climb and began to level out. A few years earlier, a major strike in the British engineering industry in general that lasted from July 1897 until January 1898, in the unsuccessful pursuit of an eight-hour working day, seriously affected several private locomotive builders, thereby greatly delaying construction and subsequent deliveries. This was shortly followed by a sudden and unprecedented industrial boom in 1899. These two unconnected factors combined to produce a quite exceptionally heavy demand for new locomotives that was barely satisfied even when very nearly two thousand entered service on Britain's railways during 1899–1900 alone. Of these, no fewer than 730 were inside-cylinder 0-6-0s, plus eighty outside-cylinder 2-6-0s imported from the USA, which had collectively been ordered by the Midland, Great Northern and Great Central railways in lieu of yet more of the urgently needed home-built 0-6-0s.

Even as late as October 1900, during the course of a single day, the then largest British locomotive builder, Neilson & Co. in Glasgow, logged orders from the Midland Railway for sixty, and from the Great Central for forty 0-6-0s (the GCR had initially enquired for seventy-five). For the Caledonian Railway during 1899–1900 all three of the then still-rival Glasgow builders, together with the CR's own St Rollox Works, were almost simultaneously building McIntosh 812 class 0-6-0s. Whereas the Caledonian Railway put a total of 159 new locomotives of various classes into traffic during 1899–1900, its corresponding total for 1901–02 was no more than eleven.

After 1900, British total locomotive stock continued to increase, but at a significantly much slower rate than hitherto. Despite total train mileage being on the increase once again, the reported locomotive grand total actually fell slightly during 1905–06. This was mainly as a result of heavy scrapping on the LNWR, which had begun to scrap Francis Webb's compound 2-2-2-0s in late 1903 within months of his departure, but before they could be adequately replaced, while the Midland sold fifty old Matthew Kirtley 0-6-0s to Italy. Generally speaking, new locomotive construction increasingly amounted to one-for-one replacement, albeit with more up to date and usually more powerful engines, rather than continuing to expand in numbers as traffic in terms of train mileage levelled off. As a result, requirements now increasingly fell within the capacity of the larger railways' own workshops.

Overall, there was a long-term and surprisingly consistent correlation between total locomotive stock numbers and total revenue-earning (loaded) train mileage. An unusually early figure is fortunately available for the nascent LNWR during 1849, when its 457 locomotives averaged 16,482 miles apiece, while in 1913 on the LNWR, now with 3,084 engines, the average was almost identical at 16,527. Also in 1913 on the fourteen major railways combined, a highly eclectic mix of 21,551 locomotives averaged 16,446 miles apiece. Remarkably, in 1949 on the newly formed British Railways as a whole with a total steam locomotive stock of 19,790, the average train mileage per (steam) locomotive at 16,872 was scarcely changed from that on the LNWR precisely a century earlier.[2]

---

[1] Atkins, P., *The Golden Age of Steam Locomotive Building*, Atlantic Transport Publishers, 1999, p. 125.

[2] Atkins, P., British steam locomotive demography, *Back Track*, February 2001, pp.71–7.

Barry Railway Class L 0-6-4T No. 143, one of ten such non-superheated locomotives built by R & W Hawthorn Leslie & Co. in Newcastle upon Tyne in 1914. Despite the considerable popularity of the 0-6-2T in South Wales, only the Barry Railway went one further and ordered new 0-6-4Ts. The Barry ordered no further locomotives after 1914, but did have late plans for an equivalent superheated outside-cylinder 2-6-2T. *Former Ian Allan Archive*

## Table 3 Locomotive production for British railways, 1895-1916 (estimated)

| Year | Total loaded train mileage (million) | Reported total locomotive stock at 31 Dec | Estimated new locomotives built by: *Rly Works* /contractors + [imports] | Total new locos to traffic[1] | Estimated *total* loco production by major British builders |
|---|---|---|---|---|---|
| 1895 | 323 | 17,871 | *510* / 108 | 618 | 441 |
| 1896 | 339 | 18,159 | *590* / 180 | 770 | 763 |
| 1897 | 351 | 18,675 | *595* / 211 | 806 | 656 |
| 1898 | 364 | 19,005 | *596* / 102 + [1] | 699 | 828 |
| 1899 | 379 | 19,624 | *595* / 343 + [55] | 993 | 775 |
| 1900 | 385 | 20,325 | *575* / 366 + [36] | 977 | 803 |
| **1901** | 382 | 20,811 | *457* / 261 | 718 | 854 |
| **1902** | 382 | 21,220 | *477* / 151 | 628 | 779 |
| **1903** | 371 | 21,445 | *456* / 104 + [1] [2] | 561 | 845 |
| **1904** | 374 | 21,586 | *413* / 115 | 528 | 851 |
| **1905** | 378 | 21,280 | *405* / 50 + [2] [2] | 457 | 900 |
| **1906** | 384 | 21,238 | *458* / 53 | 511 | 978 |
| **1907** | 393 | 21,488 | *499* / 75 | 574 | 1032 |
| **1908** | 386 | 21,564 | *384* / 108 | 492 | 928 |
| **1909** | 381 | 21,747 | *365* / 72 | 437 | 850 |
| **1910** | 384 | 21,808 | *360* / 89 | 449 | 683 |
| **1911** | 389 | 21,849 | *320* / 46 | 366 | 738 |
| **1912** | 374 | 21,969 | *348* / 124 | 472 | 731 |
| **1913** | 387 | 23,664 [3] | *367* / 66 | 433 | 898 |
| **1914** | N/A | N/A | *330* / 59 + [10] | 399 | 840 |
| 1915 | N/A | N/A | *132* / 66 | 198 | 601 |
| 1916 | N/A | N/A | *147* / 53 | 200 | 350 |

[1] Railmotor locomotives built as separate units have been omitted from the annual new-build totals. A total of at least sixty of these was built for the ADR, GNR, GSWR, LYR, LSWR, NSR, Rhymney Rly, SECR and TVR between 1905 and 1911. Most had very short working lives, although a single LYR example did continue to operate until 1948.

[2] Compound 4-4-2s imported from France by the GWR for purely evaluation purposes.

[3] Revised reporting methods in 1913: total locomotive stock now included duplicate stock that had not previously reported by some railways.

On the basis of Table 3 figures, of the total of seven thousand new locomotives estimated to have been put into service on Britain's railways between 1901 and 1914, 80 per cent of these were built by the railways themselves. The balance, however, amounted to only around 10 per cent of the combined production of the larger British locomotive builders during the same period, which was otherwise mainly destined for railways in India, South America, South Africa and Australia. This prominently included the new (and supposedly) highly standardised 4-4-0, 4-4-2, 0-6-0, 4-6-0 and 2-8-0 designs for the railways in India inaugurated in 1902 to fend off American competition, as had been experienced during 1899–1900. The British locomotive industry had already been piqued for many years by how disappointingly little it was patronised by Britain's own railways, except at times of high or sudden demand, due to the extraordinary extent to which many of these had increasingly come to build their locomotives in their own workshops since the early 1840s.[3] Back in 1875 several builders had banded together to form the Locomotive Manufacturers Association, in order to bring a legal action against the London & North Western Railway, which it had been discovered had recently built locomotives at Crewe for the Lancashire & Yorkshire Railway. This had been partly in anticipation of their proposed forthcoming mutual amalgamation, which in the event was rejected by the government. (Francis Webb is also reputed, however, to have made the offer that Crewe Works was prepared to build locomotives for anyone.)

Such facilities were originally established by the railways simply to *repair* their locomotives, but from as early as 1843 certain railways began to *construct* these for themselves as well, such that by 1870 more than 50 per cent of their locomotives, when considered overall, were being built in house. Although the relative proportions fluctuated somewhat from year to year depending on demand, this trend steadily increased, although there were numerous small railways that on account of their size were inevitably dependent on the private builders. The latter were fortunately somewhat compensated by the rapidly increasing requirements of the expanding railway systems in Britain's empire, dominions, and regions of financial influence, particularly Argentina, at a time when Britain was still perceived to be 'The Workshop of the World'. However, numerically speaking, locomotive construction in British railway workshops reached its zenith in 1892 (when it has been estimated by the author to have totalled 621 new engines). Simple statistical analysis suggests that, notwithstanding the subsequent dramatic events of 1899–1900, in reality around 1892 may well have constituted the hidden high point regarding the collective locomotive requirements of Britain's railways. There almost immediately followed a six-week strike by the miners in 1893, during which their combined train mileage fell below that of the preceding year for the first time on record. This also seriously affected the profitability of the railways for the next year or so, not least that of the LNWR.[4] Interestingly, this very period also coincided with Britain perceptibly beginning to lose its primacy as the world's leading industrial power, yielding to both the United States and imperial Germany.

## Table 4 The inauguration of in-house locomotive building by the leading British railway companies, or their direct forerunners, 1843–69

| Rly | Works | Date | Rly | Works | Date |
|------|-------|------|------|-------|------|
| LNWR | Crewe | 1843 | NER | Gateshead (–1910†) | 1849 |
| LSWR | Nine Elms (London) (–1908†) | 1843 | | Darlington, 1864– | |
| | c/a Eastleigh (So'ton), 1910– | | MR | Derby | 1851 |
| NBR | Cowlairs (Glasgow) | 1844 | GER | Stratford (E. London) | 1851 |
| GWR | Swindon | 1846 | LBSCR | Brighton | 1852 |
| | Wolverhampton, 1859–1908† | | SER | Ashford (Kent) | 1853 |
| CR | Greenock (–1855†) c/a | 1846 | GSWR | Kilmarnock | 1857 |
| | St Rollox (Glasgow), 1854- | | GCR* | Gorton (Manchester) | 1857 |
| LYR | Miles Platting (Manchester.) (–1881†) | 1847 | GNR | Doncaster | 1867 |
| | c/a Horwich (Bolton), 1889– | | LCDR | Longhedge (Battersea) (–1904†) | 1869 |

† ceased locomotive production * as Manchester Sheffield & Lincolnshire Railway.

---

[3] Atkins, P., Do it yourself locomotive building: a peculiar British custom, Railway Archive, No. 27, June 2010, pp 45-55, and No. 28, September 2010, pp. 45-52.

[4] Reed, B., *Crewe Locomotive Works and its Men*, David & Charles, 1982, p. 123. This delayed the emergence of the first three-cylinder compound 0-8-0s, and eight of the first ten compound 2-2-2-2s by about a year. The LNWR dividend also fell from 7 to 5 per cent.

## Table 5 Estimated total locomotive production by and for the fourteen major British railway companies 1901–14, also showing the percentage of these built in railway's own workshops

| Railway | By own works | By contractors | Total | % by own works |
|---|---|---|---|---|
| GWR | 885 | 3 | 888 | 99.7 |
| LNWR | 1134 | 0 | 1134 | 100.0 |
| NER | 523 | 70 | 593 | 88.2 |
| MR | 262 | 145 | 407 | 64.4 |
| NBR | 150 | 189 | 339 | 44.2 |
| GER | 505 | 0 | 505 | 100.0 |
| CR | 268 | 0 | 268 | 100.0 |
| LSWR | 206 | 1 | 207 | 99.5 |
| GNR | 515 | 14 | 529 | 97.4 |
| GCR | 292 | 361 | 653 | 44.7 |
| SECR | 177 | 76 | 253 | 70.0 |
| LYR | 459 | 0 | 459 | 100.0 |
| GSWR | 60 | 60 | 120 | 50.0 |
| LBSCR | 166 | 43 | 209 | 79.4 |
| Other* | 89 | 434 | 523 | 17.0 |
| **Total** | **5691** | **1397** | **7088** | **80.3** |

*During this period locomotives were also built in house by the Cambrian (at Oswestry), North Staffordshire (Stoke), Highland (Inverness), Great North of Scotland (Inverurie), Midland & Great Northern Joint (Melton Constable), North London (Bow), and also by the Midland Railway at Derby for the Somerset & Dorset Joint. Most prominent in this respect was the NSR, which built fifty-three locomotives at Stoke, while obtaining only five from contractors.

The Great Eastern Railway, which could be regarded as an average major railway company, built all of its locomotives in house between 1885 and 1919. The table below gives the distribution of types built at Stratford Works between 1901 and 1914. It will be noted that, as was frequently the case elsewhere, locomotives were very often built in batches of ten, as ten pairs of plate frames could be cut, or 'slotted', in a single operation, although completion did not invariably fall tidily within a given calendar year. The particular dominance of four-coupled tank engines, which were mainly intended for suburban service, will be noted.

Notwithstanding the heady events of 1899–1900, it was in 1907, as it was also in the United States,

## Table 6 Locomotive production at Stratford Works, GER, 1901–14, by locomotive wheel arrangement

| Year | 4-4-0 | 0-6-0 | 4-6-0 | 2-4-2T | 0-6-0T | Other | Total |
|---|---|---|---|---|---|---|---|
| 1901 | 10 | 25 | | | 10 | 6 | 51 |
| 1902 | 10 | 15 | | 15 | | 21 | 61 |
| 1903 | 14 | 5 | | 10 | 2 T | 2 | 33 |
| 1904 | 6 | | | 10 | 20 | | 36 |
| 1905 | | 10 | | 16 | | | 26 |
| 1906 | 8 | 20 | | 14 | | | 42 |
| 1907 | 2 | | | 39 | | | 41 |
| 1908 | 10 | | | 15 | 3 T | | 28 |
| 1909 | 10 | | | 25 | | | 35 |
| 1910 | 20 | 8 | | 2 | | | 30 |
| 1911 | 10 | 2 | 1 | 17 | 1 T | | 31 |
| 1912 | | 19 | 4 | 3 | 10 | | 36 |
| 1913 | | 11 | 14 | | 7 | 1 | 33 |
| 1914 | | | 13 | | 3 + 3T | 4 | 23 |
| **Total** | **100** | **115** | **32** | **166** | **59** | **34** | **505** |

T = 0-6-0T tram engine

Hull & Barnsley Railway Class A 0-8-0 No. 121, built by the Yorkshire Engine Co. in 1907. These were the largest locomotives to be built to the design of any member of the Stirling family. *Former Ian Allan Archive*

and collectively in the world as a whole, that steam locomotive construction in Britain, by railway works and by the private builders combined, including for industry and export, collectively attained its all-time peak. The estimated total production in Britain was 1,840 engines from all sources. This was, nevertheless only about one quarter of the corresponding total locomotive production in the United States. This was dominated by the activities of the Baldwin Locomotive Works, which alone produced 2,655 locomotives during that year. By way of comparison, in 1907 the *combined* production of the larger commercial British locomotive builders, passed above the one thousand mark for the first and only time. However, no more than sixty of these locomotives, or scarcely 6 per cent of their collective production, were destined for the home railways, which built an estimated five hundred new engines themselves. The balance of fifteen locomotives supplied by contractors to the home railways were all 0-8-0s built for the Hull & Barnsley Railway by the Yorkshire Engine Company. No doubt in a bid to undercut its competitors, this builder quoted £3,560 per engine, and in doing so made a very serious financial loss on them. It then quickly took an order for five more, this time quoted at £4,260 each with the intention of recouping its losses on the previous batch.[5]

Before 1907 had ended there was a financial panic in the USA that had worldwide repercussions, and in Great Britain the Locomotive Manufacturers Association noted a worrying dearth in orders coming in from overseas. It therefore resolved to mount a semi-covert operation to try and persuade the major British railways to place more orders with its members. During January and February 1908 a short article followed by three anonymous letters, variously signed 'Accountant (Retired)', 'ex-Locomotive Superintendent' and 'Consulting Engineer' appeared in the pages of *The Railway Gazette*.[6, 7, 8, 9] Each of these robustly refuted the railway companies' claims that they could build locomotives more cheaply than contractors, and plainly stated that this was a job for the manufacturers. The correspondents conveniently overlooked the fact that commercial manufacturers would normally seek a profit margin of say, 10 per cent, and would often incur additional development costs, for example for pattern making, and the need to make flanging blocks for boilers, for what was quite often only a short production run. Finally, there could also be delivery charges, to be met by the builder, included in the final invoiced cost.

On occasion delivery routing details from builder to customer could be surprisingly complex, such as this extreme example taken from the Robert Stephenson & Co. (Darlington) order book. Thus, the locomotive superintendent of the Rhymney Railway, when ordering ten new 0-6-2Ts in December 1909, stipulated that the first five engines should be delivered via Normanton, Manchester, Warrington, Chester and Hengoed so that the *GWR* should get the maximum mileage out of them. However, of the remainder two were to be sent via the Great Central (presumably as far as Wrexham), two by the

=5 Vernon, T., *Yorkshire Engine Company, Sheffield's own locomotive manufacturer*, The History Press, 2008, p. 53.

6 Should railway companies build their own locomotives? *The Railway Gazette*, 3 January 1908, p. 4.

7 *Ibid*, 24 January 1908, letter, p. 73.

8 *Ibid*, 31 January 1908, letter, p. 98.

9 *Ibid*, 14 February 1908, letter, p. 148.

LNWR, and one via the Midland, 'so that the *Rhymney* could get the maximum mileage from each of these companies'. The rationale underlying these very precise instructions is unclear.

It was fairly rare to find instances of the simultaneous construction of a given locomotive type by a railway works and a contractor, and for which building costs for both are also on record. However, the prices quoted below would nevertheless seem to support the railway companies' viewpoint on this matter, although their methods of costing, particularly as regards attributing overhead costs, were not necessarily strictly comparable with those of the commercial builders.

There is no evidence that this exercise by the LMA produced any positive results, although shortly afterwards neighbouring Beyer, Peacock & Co. produced for the Great Central a possibly unsolicited scheme for a large 4-4-0 with inside Walschaerts valve gear.[10] Although no order ensued, this proposal conceivably pointed the way to the Robinson large superheated Director class 4-4-0s, that were later built by the GCR itself at Gorton Works in 1913.

With regard to the division of labour of the commercially built locomotives, the contribution made by two of the oldest established builders, the Vulcan Foundry and Robert Stephenson & Co. (which had recently reconstituted and removed from Newcastle to Darlington), was surprisingly modest at around only seventy each, which was actually equalled by the small Yorkshire Engine Company, due to a surprising order for thirty 4-4-2Ts from the North British Railway. For its part Beyer, Peacock's tally was 150, but unsurprisingly the lion's share was built in Glasgow, by the recently formed North British Locomotive Company and its pre-1903 constituents. NBL was well patronised by its namesake the North British Railway, and also by the Glasgow & South Western and Highland railways. It has been suggested that this was not entirely due to innate patriotism but more prosaically because of its lower prices, resulting from the lower wages that were paid north of the border. After its formation the North British Locomotive Company claimed to be the largest locomotive builder in Europe, with its three works in Glasgow having an avowed combined capacity to build 700 locomotives per annum. NBL never actually exceeded 530, which it achieved in 1905, whereas on several occasions the annual production of Henschel & Sohn at Kassel in Germany comfortably exceeded one thousand locomotives!

## Table 7 Comparative costs of railway-built and contractor-built locomotives

| Year | Railway & type | Railway works cost | Contractor cost |
|------|----------------|--------------------|-----------------| 
| 1899-1900 | CR 0-6-0 | £2,189–£2,274 | £2,985–£3,100 |
| 1903 | GCR 4-4-2T | £1,866 | £2,770 |
| 1906 | GCR 0-6-0 | £2,406 | £3,300 |
| 1908 | NER 0-6-0 | £2,986 | £3,500–£3,550 |

North British Railway 4-4-2T built in Sheffield by the Yorkshire Engine Company, one of thirty delivered between 1911 and 1913. *Former Ian Allan Archive*

[10] Atkins, P., Robinson Great Central locomotives, a post script, *Back Track*, October 2010, pp. 634–5.

During the late Edwardian period there were indications of some discontent even in Scotland with NBL. In 1910 the NBR directors had considered a quotation obtained from NBL to build six 4-4-0s to be unacceptably high.[11] As a direct consequence, the NBR went on to build these locomotives in house, and during the winter of 1910–11 it placed orders in England, not only for thirty 4-4-2 tank engines with the Yorkshire Engine Company in Sheffield as mentioned above, but also for six large 4-4-2 tender engines from Robert Stephenson & Co. in Darlington. Ironically, only six months earlier the North Eastern Railway, whose locomotive headquarters was situated in Darlington, had ordered twenty 4-4-2s from NBL in Glasgow!

NBL, and previously its pre-1903 constituents, had become accustomed to receiving orders from the Highland Railway. However, in September 1914, after the declaration of war against Germany, as a result of which the Highland Railway attained new found national strategic importance, it urgently sought quotations for six recently designed superheated 4-6-0s, of the would-be River class. However, the HR rejected NBL's quotation in favour of that from a smaller builder in northern England, R & W Hawthorn Leslie & Co. in Newcastle on Tyne. Despite the debacle that ensued (see Chapter 10), Hawthorn Leslie went on to receive further five separate orders from the HR totalling eighteen locomotives of new design up to 1919. In contrast, after 1914 NBL received only two orders totalling six locomotives of existing design from the HR, of which the condition of the three 4-6-0s on their delivery in 1917 for some reason evidently gave cause for concern.

Undeterred, the Caledonian Railway placed its first 'outside' orders for new locomotives for the first time since 1899 with NBL in 1915, which also by then was receiving orders from the NBR once again. The Glasgow & South Western Railway had never wavered, and NBL's crowning glory, as far as the independent Scottish railways were concerned, would be building the six magnificent Robert Whitelegg 4-6-4Ts for the GSWR in the spring of 1922, on the very eve of the forthcoming railway amalgamations.

During 1901–14, the Great Central Railway obtained the greatest number of locomotives from the commercial locomotive building industry, but it did not invariably favour its next door neighbour in Gorton, Beyer, Peacock & Co., with whom at times it competed to recruit local skilled labour. There were, however, at least three occasions when the GCR contracted BP to *design* as well as build some new locomotives for it, which was unusual as far as the larger British railways were concerned. As with the 4-4-0 mentioned above, its chief locomotive designer, Carl Schobelt, was usually involved. This had first occurred in early 1903 when Schobelt schemed alternative 4-4-2 and 4-6-0 express passenger engines. Two 4-4-2s and two 4-6-0s (with each pair having slightly differing cylinder diameters, 19 and 19½in, for trial purposes), were completed by Beyer, Peacock for the GCR

at the end of the year, in order that a direct comparison could be made of otherwise identical four- and six-coupled locomotives. Of these, the 4-4-2 was very quickly selected for further construction. Popularly dubbed the 'Jersey Lily', it was widely considered to be the most handsome locomotive of its period, its external styling, not least the all-important chimney profile, although this was closely patterned on the Great Central Class 8 4-6-0 recently built in Glasgow by Neilson, Reid & Co. The styling of the Great Central 4-4-2s, and the corresponding 4-6-0s, was nevertheless entirely in the elegant tradition established more than half a century earlier in Manchester by the German-born architect Carl Beyer.

Beyer had forsaken his country and profession at the early age of 20 in 1834 to become a draughtsman in Manchester with Sharp, Roberts & Co., the forerunner of Sharp, Stewart & Co., which was then primarily concerned with manufacturing textile machinery, before moving into the exciting new field of locomotive construction. Beyer then progressed to enter into partnership with Richard Peacock (born in 1820), the remarkably young locomotive superintendent of the ancestor of what later became the Great Central Railway. Peacock had already in the 1840s established the railway's locomotive repair works at Gorton, before leaving this to co-found Beyer, Peacock & Co. on an adjacent site there also in the early 1850s. Beyer, Peacock & Co (aka The Gorton Foundry) completed its first locomotives (for the GWR) in 1855, which was followed by the first from Gorton Works directly opposite just two years later in 1857 for what by then had become the Manchester Sheffield & Lincolnshire Railway.

In June 1907, Carl Schobelt also schemed the massive three-cylinder 0-8-4Ts for the new GC concentration yard at Wath upon Dearne, the first of which was delivered only six months later during the following December. In early 1906 Schobelt's colleague, Samuel Jackson, had designed with remarkable speed, likewise working in some existing standard major components that had already been designed and made by the builder, a new fast goods 4-6-0 for the GCR (Class 8G). Evidence for this is that merely four weeks elapsed between the date of Jackson's initial diagram and the completion of the general arrangement drawing. Delivery, however, began one month late, in September 1906, when the engines came out at 3½ tons above the original weight estimate. For reasons that are now unclear, these ten relatively little known engines had evidently been required with some urgency.

For sheer speed of delivery, however, the construction by Beyer, Peacock of five production 4-4-2s in 1904 was rarely matched by the British locomotive industry. Ordered on 18 March, these locomotives were delivered only four months later during the following July. The boilers for these all having been tested during June, each engine later spent a fortnight in the paint shop, as stipulated in the GCR specification.[12] BP would already have had the necessary

---

[11] Thomas, J., *The North British Atlantics*, David & Charles, 1972, p. 105.
[12] A smart piece of locomotive building, *Engineering*, 5 August 1904, p. 191.

The first North Eastern Railway Class V 4-4-2, No. 532, probably when new. In the Class V1 'repeats' later built at Darlington Works in 1910, the aesthetics were somewhat improved by making the coupled wheel splashers narrower and mounting a cylindrical smokebox on a cast saddle. All twenty V/V1 engines were later superheated and remained in service until the 1940s. *Former Ian Allan Archive*

drawings and wheel and cylinder patterns immediately to hand as they had been produced for the very recent pilot order. However, on these repeats the fireboxes were made somewhat deeper than before as was permitted by the 'Atlantic' wheel configuration, when compared with a shallower arrangement designed earlier such that it would suit both the pilot 4-4-2s and 4-6-0s.

It was normally reckoned to take three months to make new cylinder patterns, and a good illustration of this sometimes critical factor was provided by the North Eastern Railway Class V 4-4-2s built at Gateshead Works, which lacked foundry facilities. Twenty engines (later reduced to ten) were authorised by the NER Locomotive Committee on 29 January 1903. The drawing for the cylinders, the design of which was usually a high priority for an entirely new locomotive design, was dated only eight days later. The cylinders would then have been ordered very promptly from Kitson & Co., who supplied the first pair of castings on 15 June. Sir James Kitson, later Lord Airedale, was for many years a director of the NER, and during 1904–06 was the chairman of its locomotive committee. Casting cylinders was a particular speciality of his eponymous locomotive building enterprise in Leeds that, together with the fact certain other NER directors also owned iron foundries, was the reason why neither Gateshead nor Darlington Works were provided with such facilities.[13] The first 4-4-2 boiler was completed on 4 September at Gateshead, from whence the prototype locomotive, NER No. 532, made its maiden trip up to York on 3 November 1903, drawing a saloon conveying the NER's chief mechanical engineer, Wilson Worsdell. On the strength of

observations made on this run, modifications to raise the cab roof, in order to improve forward visibility from the footplate, were put in hard immediately thereafter.

For the later and more complicated NER four-cylinder compound 4-4-2 (1906), a very detailed breakdown of the actual building cost has survived. Total cost was £4,658, inclusive of materials, wages and overheads. The price of the individual *major* components were as follows: *Engine*, boiler £955, wheels £443, cylinders (cast by R. Stephenson & Co.) £313, frames £160, and valve gear £80. *Tender*, tank £126, wheels £125, and frames £58.[14] These now seemingly relatively paltry sums need to be multiplied by a factor of approximately 120 in order to equate to 2018 price levels!

Usually 'outside' locomotive orders were put out to competitive tender. Printed specifications, accompanied by major drawings, were sent out to contractors who would then quote a price, which was often based on their calculated 'dry' or empty weight not given in the specification, also stated, and the anticipated date of delivery. Successful selection would primarily be determined on price, but quick delivery (not always honoured) might also be a deciding factor. The entire process was often accomplished within just fourteen days. In March 1905, the London Brighton & South Coast Railway invited bids for the supply of five 4-4-2 tender locomotives, which were to be based closely on the Ivatt Great Northern Railway large-boilered 4-4-2s with their wide fireboxes. Doncaster working drawings for these were used as the basis, amended where necessary in red ink. The quotations received were as follows:

[13] Heppell, G., *North Eastern Locomotives: A Draughtsman's Life*, North Eastern Railway Association, 2012, p. 8.

[14] Hoole, K., *The North-Eastern Atlantics*, Roundhouse Books, 1965, pp. 22–3.

## Table 8 Quotations received to build 5 4-4-2s for the LBSCR, March 1905

| Builder | Cost per engine | Promised delivery dates |
|---|---|---|
| Kitson & Co. | £3,950 | 1st on 22 August, then one per week |
| Robert Stephenson & Co. | £4,132 | 1st on 31 December, then two per week |
| North British Loco. Co. | £4,185 | 1st on 19 August, last on 31 August |
| Yorkshire Engine Co. | £4,200 | 1st on 31 December, then two per week |
| Vulcan Foundry Ltd | £4,410 | 1st on 15 January 1906 |
| Beyer, Peacock & Co. | £4,674 | 1st on 12 October, then one per week |

The bid received from the small Yorkshire Engine Company in Sheffield was surprising, but in the event Kitson & Co. with the cheapest offer was awarded the order. However, far from delivering the first engine in late August, this did not actually reach Brighton until early December. Such delays on the part of locomotive manufacturers, despite contractual agreements, were by no means unusual. The original contract in this instance would not appear to have included any penalty clause for late delivery, i.e. stipulating a reduction in payment, as was frequently the case. The engines were delivered in grey primer, and following trials they were to have been painted at Brighton Works in Stroudley yellow livery. However, they were actually finished in a new and cheaper umber (brown) livery, which allegedly achieved a *saving* of £50 per locomotive. Rather strangely, the cost of painting the NER compound 4-4-2 referred to earlier, a few months later at Gateshead, only amounted to £47 *in total*. Back in 1885, in response to criticism from Francis Webb, who famously painted all his locomotives black, William Stroudley himself had retorted that the actual colour that a locomotive was painted had little bearing on the cost of that particular operation, although in his case this also involved some elaborate lining out.

Like the Great Central, the Glasgow & South Western Railway was also heavily reliant on the private locomotive builders in addition to its own resources. The North British Locomotive Company (NBL) delivered ten express passenger 4-6-0s during May–June 1903, actually ordered from Sharp, Stewart & Co., and seven similar additional 4-6-0s were later built by the GSWR at its Kilmarnock Works during 1910–11. These were soon followed by a pair of superheated developments from NBL in July 1911, which showed a considerable improvement in their performance over that of the earlier non-superheated engines. A little-remarked but advanced feature of these two engines was the employment of direct outside steam pipes to the outside cylinders for the first time in a British locomotive. Even the GWR would not follow suit in this respect for another ten years.

In 1913 NBL built several unusually large and heavy coal-eating non-superheated inside cylinder 4-4-0s and 0-6-0s for the GSWR, shortly after Peter Drummond had succeeded James Manson as its locomotive superintendent. The handsome 4-4-0 was clearly closely based on Drummond's elder brother Dugald's very recent D15 class on the LSWR, but the similarly styled if rather ponderous 0-6-0 had no English counterpart, since the LSWR had built no 0-6-0s since 1897.

As earlier indicated, between 1901 and 1914 several of the major British railway companies were entirely independent of the commercial locomotive builders. Although the newly established South Eastern & Chatham confederation was one of three railways that in 1901 could boast of *two* locomotive-building facilities, it was not one of these. Locomotive construction would cease in 1904 at Longhedge Works in Battersea (from its London, Chatham & Dover Railway component) and thence be concentrated at Ashford (South Eastern). As from 1909 the Great Western would demote Wolverhampton to carrying out locomotive repairs, with all new construction thereafter confined to Swindon. In 1910 the North Eastern built its last locomotives at Gateshead, having recently expanded its workshop facilities in Darlington. The distinctly reduced requirement for new locomotives after 1900 could well have been a factor in these rationalisations. Also in 1910, after several decades of ongoing debate, the London & South Western finally closed down its Nine Elms Works in London completely, and transferred all locomotive building and repairs to new facilities at Eastleigh (Southampton), were it already had established carriage and wagon-building facilities. Likewise, at this time the London Brighton & South Coast Railway was considering establishing a new locomotive works at Lancing, where its own rolling stock building and repair works was situated. In fact, locomotive construction would continue at Brighton until 1957.

During 1913–14 the larger locomotive builders were quite busy once again, due to large orders from overseas, especially India. In 1914 only twenty-two locomotives, 4-4-0s for the South Eastern & Chatham Railway, were built 'outside' for any of the fourteen major railways in Britain. Twelve of these had been ordered in December 1913 from Beyer, Peacock & Co. for delivery by June 1914. A further ten were to be built by Ashford Works but, for speed of delivery, in February 1914 the SECR announced that these had been ordered from Borsig A. G. in Berlin. It was stated that these were wanted in time for the start of the summer traffic in May, which the German enterprise had promised, and which British builders could not undertake, while emphatically

South Eastern & Chatham Railway Class L 4-4-0 No. 779, built in Berlin in 1914. Appropriately, the German-built engines differed from their British-built counterparts in being fitted with Schmidt rather than Robinson superheaters. They were fully painted in SECR green livery (as per specification detailed in Chapter 14) after their delivery and full assembly at Ashford Works. *Former Ian Allan Archive*

stating that price and quality had not been an issue.[15] The railway was criticised for leaving the order too late, which could otherwise have been handled by any one of several British builders. A representative from an unspecified 'Glasgow undertaking', but in other words NBL, was quoted as saying that (despite having *three* works at its disposal) it would have been prepared to deliver during July–August.[16]

The first five 4-4-0s were landed at Dover in dismantled form on 19 May, and together with the remaining five these were erected at Ashford Works by Borsig's fitters, who had only just returned to their homeland when war between Britain and Germany was declared on 4 August. It had indeed been a remarkable achievement on Borsig's part, involving twice as many locomotives than those that had been built by Beyer, Peacock ten years earlier for the GCR, and over a similar time scale. Furthermore, it would have required the fabrication in Berlin of new wooden patterns for wheels and cylinders, etc, to unfamiliar imperial dimensions.

The ten Borsig engines duly entered traffic on the SECR during June–July, the first barely five months after being ordered, and were then directly followed by the Beyer, Peacock twelve, delivered complete between August and October 1914. The last of these by

comparison arrived *ten months* after it had been ordered. (During 1914, BP built only ninety-nine locomotives, compared to its all-time peak of 152 only six years earlier.) Six 0-6-2Ts, which had also been ordered from another German locomotive builder by the Taff Vale Railway, were much less advanced at this point, and were thereupon cancelled. These were built later on in Glasgow by NBL. The German-built SECR 4-4-0s were eventually duly paid for, with interest, eighteen months after the Armistice, in May 1920.

The war would change everything, most particularly on the LNWR. Upon its declaration, Crewe Works immediately suspended all new locomotive construction, and no new locomotives emerged from there between November 1914 and April 1915, while repaired locomotives were returned to traffic not fully repainted and un-varnished.[17] At the end of March 1915 copies of the general arrangement drawings for the 'Prince of Wales' 4-6-0 and G1 0-8-0 were sent to North British, Vulcan Foundry, Kitson & Co., and Beyer, Peacock, with invitations to quote to build twenty of each. NBL landed the 4-6-0 order, which did not include tenders. Although delivery of the 4-6-0s had been contracted to begin in October 1915, the first engines did not arrive at Crewe until early March 1916.

[15] German engines for British railways, *The Railway Gazette*, 20 February 1914, p. 20.

[16] British v. foreign locomotives, *The Railway Times*, 7 March 1914, p. 208.

[17] *The Locomotive Magazine*, September 1914, p. 235.

Meanwhile, the 0-8-0 order had been awarded to Beyer, Peacock & Co. In this instance Crewe appears to have already made a start on these engines before the war had begun, and was able to pass some ready-made components on to the Gorton Foundry. In the event the latter made no progress at all on the contract owing to it having to prioritise munitions manufacture as demanded by the government. This resulted in a parliamentary question in October 1917 concerning the total of forty-five unfinished locomotives that were languishing at the Gorton Foundry.[18] As a result only the following month the LNWR order was taken back again by Crewe, which turned the engines out during 1918 nearly three years late.[19] Beyer, Peacock also passed on an order placed in January 1915 by the GNR for twenty 2-6-0s to NBL.[20] However, it did not finally deliver five small 0-6-0s that had been ordered in January 1915 by the Cambrian Railways until between October 1918 and June 1919.[21]

After the war, in 1919 the LNWR placed a very large order for ninety Prince of Wales 4-6-0s, and for sixty tenders, with William Beardmore & Co. in Glasgow, at a total cost of £800,000 (the equivalent of £40 million at 2018 prices). Beardmore, like Armstrong Whitworth & Co in Newcastle, had just entered the locomotive-building industry in the immediate aftermath of the Armistice. Boilers for new post-war batches of Claughton 4-6-0s building at Crewe were outsourced to Vickers in Barrow and Nasmyth Wilson & Co. in Patricroft, while other boilers were similarly ordered from NBL for fitting to the new G2 0-8-0s that it was also constructing.

At this time the also hitherto independent Great Western Railway ordered fifty Churchward mixed-traffic 2-6-0s from Robert Stephenson & Co., which were the only *tender* locomotives ever to be ordered 'outside' by the GWR in the 20th century. In the event only thirty-five of these engines were actually assembled in Darlington, with the remaining fifteen sets of parts being despatched to and erected at Swindon Works, at significantly reduced cost to the GWR.

These substantial locomotive orders placed by the LNWR and GWR were funded by government compensation payments paid to them in recognition of their sterling efforts during the recent war. Comparable orders placed by other railways at this time were for fifty Class T2 0-8-0s by the NER with Sir W. G. Armstrong Whitworth & Co., and for fifty Class N2 0-6-2Ts by the Great Northern Railway with NBL. Perceived priorities differed very considerably, for between them they were for express passenger, mixed traffic and heavy goods tender engines, and suburban passenger tank locomotives.

Unthinkable only three years earlier, an LNWR locomotive carrying a commercial locomotive builder's works plate. An NBL official photograph of one of the twenty LNWR Prince of Wales class 4-6-0s completed in Glasgow during 1916. The builder's works plates mounted on the side of the smokebox were indeed very soon removed after the engines had entered service.

## Railway works

During 1913–14 *The Railway Magazine* ran a series of interesting well-illustrated features on British railway workshops: Kilmarnock (GSWR) January 1913, St Rollox (CR) March 1913, Brighton (LBSCR) June 1913, Horwich (LYR) August 1913, Stoke (NSR) November 1913, Derby (MR) April 1914, and Gorton (GCR) November 1914. Other similar articles would probably have followed but for the international situation. However, during the war, *The Railway Gazette* ran a highly detailed and extensively illustrated feature 'The Production of a Locomotive Valve Gear' (for the LNWR Claughton class 4-6-0) in its issues dated 27 April, 11 May, and 25 May 1917. This was soon followed by 'The Stooperdale Boiler Shop of the North-Eastern Railway at Darlington', which particularly illustrated the manufacture of new boilers for the NER Class Z 4-4-2s, in its issues dated 24 and 31 August 1917. With reference to a post-1914 locomotive design (the GNR three-cylinder 2-6-0s), *The Railway Engineer* for April 1920 fully described 'The Casting and Machining of Locomotive Cylinders'. These three articles collectively and wonderfully described the then routine, but nonetheless high-precision, procedures involved in the construction of a contemporary modern steam locomotive that took place in around a dozen railway and a similar number of commercial locomotive building workshops in the early 20th century.

[18] *The Railway Gazette*, 26 October 1917, p. 464.

[19] Atkins, P., Going 'outside', *LNWR Society Journal*, September 2001, pp. 185–90.

[20] *Locomotives of the LNER*, Part 6A, Classes J38 to K5, The Railway Correspondence & Travel Society, 1982, p. 60. The GNR 2-6-0s were delivered between June and August 1918.

[21] *The Locomotives of the Great Western Railway*, Part 10, Absorbed Engines 1922–1947, The Railway Correspondence & Travel Society, 1966, p. K71.

# 4

# 4-4-0s & 0-6-0s

Not only was a *precise* total determined for the national locomotive stock for the first time in 1913, i.e. 23,664, courtesy of the newly revised *Board of Trade Annual Railway Returns*, but it also now became possible to quantify fairly accurately the relative numbers of locomotives of the different wheel arrangements then in use. For some thirty British railway companies these were publicly tabulated for 1913 in *The Railway Year Book for 1914*. These subtotals would from then on change continually over the next fifty-five years on Britain's railways, until the demise of the steam locomotive in normal service thereon.

The wheel arrangement Top Ten, which collectively accounted for 85 per cent of the total locomotive stock as it stood on 31 December 1913, is estimated to have been as follows:

### Table 9 British locomotive wheel arrangement census, 1913

| | | | |
|---|---|---|---|
| 0-6-0 | 7,310 | 2-4-2T | 1,004 |
| 0-6-0T | 3,700 | 2-4-0 | 955 |
| 4-4-0 | 3,168 | 0-8-0 | 864 |
| 0-6-2T | 1,395 | 4-6-0 | 722 |
| 0-4-4T | 1,233 | 4-4-2T | 455 |

(See also Appendix 3 for a more detailed breakdown by major railway company)

Given the overwhelming dominance of three-wheel arrangements, in association with prominent others that had also originated in the 19th century, then at least 90 per cent of the British national locomotive stock in 1913 would have had inside cylinders. As regards tender locomotives, the dominance of the 4-4-0 passenger and 0-6-0 goods engine is unmistakeable. Approximately similar numbers of each type were built in total between 1901 and 1914, with around 1,300 4-4-0s and 1,200 0-6-0s. When combined these accounted for just over one third of all new locomotive construction by and for the British railway companies during that period. Through their sheer numbers 4-4-0s and 0-6-0s were the locomotives that earned the bread and butter.

On a number of railways 4-4-0s and 0-6-0s were often interchangeable as regards their boilers, cylinders and tenders, etc. Good examples of this logical pairing in the interests of standardisation could be found, for instance, on the Midland, Great Central, Great Eastern, and Caledonian railways, and even on the Hull & Barnsley. Almost invariably, of the two it was the 0-6-0 that lasted the longest, sometimes well into the 1960s.

The relative numbers of 4-4-0s and 0-6-0s built by each of the major railway companies varied considerably, as did the period over which they were built. Thus the LYR built no 4-4-0s after 1894, which was just when the GWR at Swindon began to build the type for the standard gauge, or narrow gauge as the GWR then continued to regard it.[1] The 4-4-0s continued to be built by some other companies through to 1922, when numerically speaking the type was about at its peak. Indeed, upon is formation in January 1923, no less than one quarter of the locomotive stock of the Southern Railway was comprised of 4-4-0s. As for 0-6-0s, the GWR ceased to build these in 1899, as did the LNWR in 1902, when the national 0-6-0 total was probably at an all-time peak of around 8,000, following the very recent construction of 700 of these during 1899–1900 alone.[2] Nevertheless, several of the other major companies continued to build 0-6-0s until 1920–22, while after 1922 each of the later Big Four railways would build them in diminishing numbers until the 1940s.

---

[1] The 4-4-0 had actually been pioneered in Britain by the GWR in 1855, on the 7ft gauge with the twelve-strong Waverley class, designed by Daniel Gooch and built by R. Stephenson & Co. The Brunel 7ft or 'Broad Gauge' was finally eliminated in May 1892.
[2] The GWR resumed 0-6-0 construction in 1930 with the Collett 2251 class, the last of which was completed in 1948.

The heaviest 4-4-0s and 0-6-0s to appear during the Edwardian period were built by the North British Locomotive Company in 1913 for the Glasgow & South Western Railway. Both classes were extremely heavy on coal, but were later followed by slightly heavier, although more successful, superheated developments of them in 1915. *Upper:* GSWR 131 class express passenger 4-4-0 No. 131 (62 tons). *Lower:* GSWR 279 class heavy goods 0-6-0 No. 300 (58 tons).

The first new British 4-4-0 design to appear in the 20th century was the celebrated South Eastern & Chatham Railway Class D 4-4-0 in February 1901. No. 729, from the initial batch delivered by Sharp, Stewart & Co. during that month, was recorded at Chislehurst in its prime at the head of a Dover express. *Former Ian Allan Archive*

MR 4-4-0 No. 999 when new. The cylindrical smokebox mounted on a cast saddle was an innovation on the Midland, while invisible was the newly patented Deeley valve gear. *Former Ian Allan Archive*

The GWR moved on from conventional single-frame 0-6-0s to building much more powerful double-frame 2-6-0s, also with inside cylinders. These consisted of the eighty Aberdare engines built between 1900 and 1907, the later examples with taper boilers, and the enigmatic and indeed bizarre Krugers, of which eight were produced between 1901 and 1903, which all had been withdrawn by late 1907![3]

## 4-4-0

The classic British inside-cylinder 4-4-0 was effectively pioneered by Samuel Johnson on the Midland Railway in 1876. His so-called 'slim boiler' 4-4-0s (boiler diameter 4ft 2in) were built over a period of twenty-five years until 1901 with incremental changes, which included the provision of very early piston valves from 1893. Their latter-day construction overlapped with that of the first Johnson Belpaire 4-4-0s in 1900, which while incorporating very similar machinery between the frames, had substantially larger 200lb pressure boilers of 4ft 8in diameter, and longer fireboxes with sloping grates, whose area had been increased from 21 to 25 sq ft.

Eighty Belpaire or 700 class 4-4-0s were built at Derby up to 1905, the first fifty being provided with large bogie tenders to several slightly differing designs. Although the final batch of these, completed during the winter of 1903–04, was fitted with water pick up apparatus, it was the rapid provision of water troughs on the Midland immediately thereafter that very quickly rendered these heavy and poor-riding vehicles unnecessary. Together with the initial ten that had been built for the final ten 4-2-2s in 1899–1900, these large tenders began to be converted to six-wheeled vehicles from 1908. Photographs showing eight-wheelers bearing 1907 re-numbers are uncommon, and they are believed to have become finally extinct c. 1915. The tenders of all but three of the 4-6-0s of the Midland's Scottish partner, the GSWR, were built to a very similar design to those constructed at Derby, of which the last of these remained in service until 1934.

In 1907 Richard Deeley built a solitary simple expansion equivalent of his compound 4-4-0s, numbered 999, which was fitted with his newly patented eccentric-free valve gear.[4] Its appearance was remarkably austere when compared with the Johnson slim-boilered 4-4-0s with their sweeping splashers that were still being built only a few years before. There was a unique proposal, not implemented, to letter No. 999's tender MIDLAND in full. Nine more production engines, numbered 990–8, later followed. By Midland standards these were truly rare birds, and after 1914 mainly operated between Leeds and Carlisle. All were quickly superheated, and were endowed with good cylinder design with short and straight steam ports. Although their boilers were dimensionally identical with those of the Compounds, they were not interchangeable with these because of the different position of the mud hole on the underside of the boiler barrel, which was necessary as it would otherwise have

[3] For a detailed analysis of the GWR Kruger 2-6-0s by I. Lewis, see *Back Track* for May, July, and September 2013, pp. 277–80, 432–7 and 547–9.

[4] Deeley, R. M., BP 16,372/1905 (locomotive valve gear).

been incompatible with the fitting of the Deeley valve gear. On this account the ten engines were quickly regarded as non-standard by the LMSR, which therefore began to withdraw the class as early as 1925. Updated with long-travel Walschaerts valve gear, after 1923 there might have been a case to continue to build improved 990s, rather than the more complicated three-cylinder compounds that the early LMSR initially regarded as its benchmark as regards relative locomotive efficiency.

The sweeping brass-beaded splashers of the earlier Johnson 4-4-0s, with their distinctive oiling apertures, featured on the first new 4-4-0 design to appear in the Edwardian period, the celebrated South Eastern & Chatham Class D. It is an interesting fact that its splashers were designed by James Clayton at Ashford, after he had left Beyer, Peacock & Co., and that this elegant styling could be traced back to Charles Beyer himself when at Sharp, Stewart & Co. more than fifty years earlier. The Class D was built until 1907, overlapping with the dimensionally similar Class E, which was most readily distinguished by its Belpaire firebox, and which also had 6ft 6in as opposed to 6ft 8in coupled wheels.

Although many D and E 4-4-0s were very extensively rebuilt between 1919 and 1927, Class D Nos 75 and 577, new in March 1903 and September 1906 respectively, had the distinction of being the last non-superheated 4-4-0s to remain in service on British Railways, when withdrawn in December 1956. The now preserved No. 737, which was retired only the previous month, dated back even earlier to December 1901, and so had notched up all but fifty-five years, covering an estimated 1,694,660 miles, while essentially retaining its original elegant format throughout.

In Scotland on the North British Railway, Dugald Drummond had introduced his Abbotsford class 4-4-0 in 1877, and this set the template for 265 4-4-0s that he subsequently successively produced for the NBR, the Caledonian and the London & South Western, including a small number that were also copied by his younger brother Peter on the Highland up to 1908.

On the LSWR Drummond had initially produced his C8 class in 1898, which was very similar to its NBR and CR antecedents, but which was allegedly not unduly successful despite having much in common with the very much longer-lasting 700 class 0-6-0s and M7 0-4-4Ts. The T9 4-4-0 that followed has been regarded as a classic, but after the last of these was completed in 1901, over the next seven years he built 110 4-4-0s of four new classes of a more general purpose nature at Nine Elms:

40 Class K10, 5ft 7in, 1901–02 (mixed traffic)
40 Class L11, 5ft 7in, 1903–07
10 Class S11, 6ft 1in, 1903 (passenger)
20 Class L12, 6ft 1in, 1904–05

Whereas the K10 class effectively utilised the Drummond C8/700/M7 boiler, the L11 that immediately followed used that of the T9 class. Classes S11 and L12 had 5ft

diameter boilers but the same 24 sq ft grates of the T9s. On all three classes this was achieved by pitching the coupled axles at 10ft centres and resorting to unduly narrow 2in water legs at the foundation ring, an undesirable Nine Elms/Eastleigh trait that would be perpetuated even in the Urie 4-6-0 classes, including the Southern Railway-built King Arthurs.

The classic Drummond 4-4-0 was notably further developed on the Caledonian Railway under John McIntosh from 1896 with his Dunalastair class, which inaugurated the 4ft 9in diameter boiler to British locomotive practice. The Dunalastair II and III classes that also featured this quickly followed in 1897 and 1899. The Mark IV 4-4-0 with 5ft diameter boiler did not appear until 1904, and was further updated to very good effect with superheater and piston valves in new builds from 1910. Of all the Edwardian 4-4-0 designs, the thirty-six Dunalastair IVs built between 1904 and 1914 with their bogie tenders and blue livery were particularly appealing. They formed the basis for forty-eight dimensionally similar, but less highly regarded, engines with six-wheeled tenders built by William Pickersgill, which entered traffic between 1916 and 1922.

Painted a darker shade of blue were the Great Eastern Claud Hamilton 4-4-0s, which were actually introduced in 1900, of which a Belpaire firebox version appeared late in 1903. The latter remained in regular production until early in 1911, when Stratford Works re-tooled in order to begin to build instead the new 1500 class 4-6-0. Despite the latter being designed to meet exacting weight restrictions, it is therefore something of a mystery that in January 1914 Stratford drawing office produced a diagram for a large 4-4-0, utilising the same 20in by 28in cylinders and a slightly shortened version of the 4-6-0 boiler, with no less than 21 tons reposing on each coupled axle! This must have been something more than an idle fantasy, because a few months later Stratford proposed yet another 4-4-0, this time having a slightly larger boiler, but with coupled wheels of only 5ft 8in diameter (a GER standard), which once again had an anticipated 21-ton axle load.

After an interval of no fewer than twelve years, Stratford 'posthumously', as far as the former GER was concerned, produced another ten Claud Hamilton 4-4-0s in 1923 to the original basic design, except that their boilers were increased in diameter from 4ft 9in to the 5ft 1in of the 4-6-0s. Despite also being superheated there were no increases whatsoever in their basic cylinder dimensions, boiler pressure, or grate area when compared to GER No. 1900 *Claud Hamilton* as built back in 1900. They still even retained slide valves beneath the cylinders, and although they became known as Super Clauds, they were nowhere near as 'super' as the very striking 4-4-0 proposed in January 1914 would have been!

Also in 1914, there is evidence that Nigel Gresley on the Great Northern was contemplating building a large superheated 4-4-0 with 20-ton axle load. Construction of 4-4-2s by the GNR had ceased in 1910, and these had always overshadowed the 136 unremarkable 4-4-0s that were built by Henry Ivatt at Doncaster between 1897 and 1911.

CR Dunalastair IV 4-4-0 No. 146 passes Rockcliffe with an up express, c. 1913. *H. G. Tidey/Rail Archive Stephenson*

Great Eastern Railway Claud Hamilton class 4-4-0 No. 1831, built at Stratford Works in March 1908. The first fifty engines built up to 1904 were originally oil burners, operating on the Holden system. *Former Ian Allan Archive*

The North Eastern Railway equivalent of the Great Eastern Claud was Wilson Worsdell's Class R 4-4-0 introduced in 1899, which remained in production at Gateshead until 1907. The prototype, NER No. 2011, new in August 1899, ran 139,543 miles up to 31 December 1901, and a further 64,932 miles during the six months ended 30 June 1902. During this period, double manned, it ran a lengthy daily 455-mile diagram, Newcastle–Edinburgh–Newcastle–Leeds–Newcastle, for five days a week, and then 161 miles on the Saturday, with washing out of the boiler on the Sunday, to give 2,436 miles in a week. It was considered that this level of performance had only been possible due to the provision of Smith piston valves, although it had originally been proposed to build the R class alternatively with piston valves or slide valves, as had been the case with the contemporary Class S 4-6-0 and Class T 0-8-0.

Four-cornered dynamometer car tests were held on the NER in the autumn of 1907 between York and Newcastle, which involved an R class 4-4-0, V and brand new 4CC class 4-4-2s, and an S1 4-6-0, simply to determine their respective drawbar pull on level track at 55mph. In terms of efficiency, at a typical average working speed, the older and smaller R design held up well against the larger and later ten-wheelers:

Four railway companies: tender and tank, passenger, mixed traffic and goods. Their hammer blow, which is a consequence of the normally necessary reciprocating balance weights incorporated within the coupled wheel castings, increases as *the square of the rotational speed*, and was tabulated at 6 revolutions per second. However, in many instances the values given were purely theoretical as most of the locomotives tested, with their short travel valves, etc, could only achieve around 5rps at best, for which the hammer blow would amount to $5^2/6^2$, or only 69 per cent of the Report's standard quoted 6rps value. On this account, therefore, hereafter in this book hammer blow values as given in the Bridge Stress Committee's report have been correspondingly reduced to the lower 5rps value, which makes for more realistic comparisons between different locomotive designs.

From the published tables it is apparent that the NER Class R 4-4-0 delivered the *highest* hammer blow values of any of the 175 locomotive types tested! The highest known speed recorded by an R was 82mph (5.7rps) near Ripon in 1914, when it would have amounted to 14½ tons. The R was very closely followed in this respect by the later and larger Class R1, which for its part, on account of its unprecedented static axle load, potentially could have produced the highest ever *combined* impact. However, it

## Table 10 NER express passenger locomotive dynamometer car trials, 1907

| NER loco class | Grate area sq ft | Engine wt tons | Drawbar pull | Drawbar Horsepower (DBHP)* @ 55mph | DBHP per sq ft of grate | DBHP per ton loco wt |
|---|---|---|---|---|---|---|
| R 4-4-0 | 20 | 44.5 | 1.30 | 427 | 21.3 | 9.8 |
| S1 4-6-0 | 23 | 67.1 | 1.38 | 453 | 19.7 | 6.75 |
| V 4-4-2 | 27 | 72.7 | 1.66 | 545 | 20.2 | 7.5 |
| 4CC 4-4-2 | 29 | 73.6 | 1.88 | 618 | 21.3 | 8.4 |

* Drawbar horsepower = 5.97 x drawbar pull (tons) x speed (mph).

It will be noted that for all its sophistication, the compound 4-4-2 developed no more *useful* horsepower per square foot of fire grate, and rather less per ton, than the no frills 4-4-0. The results possibly prompted the idea of developing the trusty Class R further, but not for the first time in the Edwardian era bigger did not necessarily mean better. The resultant Class R1 4-4-0 introduced just one year later, proved to be a profound disappointment. This also boasted the unprecedented axle load for that period of 21 tons, while both the R and R1 4-4-0 classes were notable for their exceptionally high hammer blow characteristics at speed.

Fifteen years later in 1923, in acknowledgement of the dramatically increasing size and weight of locomotives, the government set up the Bridge Stress Committee in order to examine the *dynamic* impact on bridges of locomotives at speed, for which hitherto the criterion had been simply their *static* axle load. The highly detailed report was published by HMSO in early 1929, tabulating data from currently extant locomotive classes from the Big

is unlikely that an R1 could have equalled the same maximum speed of the R. Fortunately, the former steel Ure Viaduct at Ripon, along with other NER structures, successfully withstood the countless transits made by Wilson Worsdell's 4-4-0s over them during the course of around fifty years.

In 1913 the Great Central Railway introduced its now superheated Director class 4-4-0 of similar proportions to the NER R1, which are both discussed in Chapter 5. In 1912 Dugald Drummond on the LSWR produced his ultimate passenger 4-4-0 design, the D15, whose long sloping firegrate equalled the 27 sq ft of the NER R1. Interestingly, it incorporated outside admission Walschaerts valve gear, which operated 10in diameter piston valves set above the 19½in cylinders. This also provided the basis for Peter Drummond's 131 class on the Glasgow & South Western Railway, which was introduced soon afterwards in 1913. But the latter was by no means a slavish copy of its English counterpart. On the GSWR engine the coupled wheel diameter was reduced from 6ft 7in to 6ft, although the

North British Railway Scott class express passenger 4-4-0 No. 895 *Rob Roy*, built in July 1909 by the North British Locomotive Company. *Former Ian Allan Archive*

boiler diameter was increased from 5ft to 5ft 4½in, firebox water tubes were omitted, although the Drummond smokebox steam drier was retained, and the internal Walschaerts valve gear was re-arranged for inside admission.

Elsewhere in Scotland, the North British Railway under William Reid rested content with 4-4-0s having fairly moderate similar proportions and something of a physical resemblance to the McIntosh Dunalastair IVs on the Caledonian. These effectively fell into two groups:

1. Those with 6ft 6in coupled wheels for express passenger work, were given names that were associated with the Walter Scott Waverley novels.

2. Those with 6ft coupled wheels were particularly intended to work express fish traffic from Aberdeen and were nameless, although those built later and named after Scottish glens were

NBR Glen class superheated mixed traffic 4-4-0 No. 258 *Glen Roy*, built at Cowlairs Works in September 1913. *Former Ian Allan Archive*

A London & North Western Railway George the Fifth class 4-4-0, No. 2271 *T. P. Bickersteth,* heads a down West Coast express near Kenton. *Former Ian Allan Archive*

extensively employed on the West Highland line. The latter were all superheated from new, while all of the other Reid 6ft and 6ft 6in 4-4-0s were eventually superheated.

The Reid 4-4-0s had been preceded on the NBR in 1903 by the Matthew Holmes 317 class 4-4-0s, which dimensionally speaking would appear to have been jointly inspired by the CR Dunalastair III and NER Class R, also adopting the latter's Smith piston valves below the cylinders. Unusually, withdrawal of these particular 4-4-0s began as early as 1922, when their non-standard boilers began to fall due for replacement.

By general consent, performance wise, the most outstanding 4-4-0s of the Edwardian era were of the Bowen Cooke LNWR George the Fifth class, introduced in 1910 as a superheated development with piston valves of the Whale Precursor of 1904. Although indicator trials took place with *George the Fifth,* unfortunately no details appear to have survived for comparison with those that were obtained from *Precursor* back in 1904, but their maximum output could well have been of the order of 1,400 IHP. Many of the Precursor class were subsequently rebuilt with superheaters and piston valves effectively to become 'Georges', whose star briefly shone very brightly. After 1914 passenger trains became slower and heavier, factors which favoured the Prince of Wales 4-6-0s, which had similarly been directly developed from the previous Experiment class, and which thereafter were built in comparatively large numbers until 1922.

Compared with other British 20th century inside-cylinder 4-4-0 classes, the LNW Precursors and Georges were unusual in being fitted with Joy valve gear. On the Georges this was arranged to produce an unusually long valve travel of 5½in, in conjunction with a variable blastpipe arrangement, which almost certainly contributed to their fine early performances.

Relatively few 4-4-0s were built for the LBSCR, no more than fifty-eight in fact. The last thirty-three of these, which comprised the Robert Billinton B4 class, was delivered between 1899 and 1902. However, they were complemented by the twenty-seven-strong I3 class 4-4-2Ts, built by Douglas Earl Marsh and Lawson Billinton, at Brighton between 1907 and 1913, twenty of which were superheated from new. These could be regarded as tank engine equivalents of the B4 class. They were readily capable of running the 51 miles between Brighton and London Victoria non-stop, for which tenders were not essential.

With regard to the smaller railways, the very extensive use of the 4-4-0 by the Great North of Scotland Railway, which completely excluded the usual complementary 0-6-0, was exceptional. At the end of 1913, out of its total locomotive stock of 117, ninety-seven engines were 4-4-0s, sixty-nine of these having inside cylinders and dating from 1884 onwards. Effectively one *basic* 4-4-0 design, introduced in 1893 with 6ft 1in diameter coupled wheels, was then built intermittently until 1921, some of them by the GNSR at Inverurie, of which the later examples were superheated. Five 4-4-0s built in 1900 by Neilson & Co. for the GNSR were instead sold to the SECR.

Midland & South Western Junction Railway 4-4-0 No. 1, built by the North British Locomotive Company in 1905. *Rail Archive Stephenson*

## A quartet of 4-4-0s

In England, the North Staffordshire Railway built four 4-4-0s at Stoke in 1910 specifically to work its summertime holiday workings between Derby, Stoke, and Llandudno. A superheated fifth engine, which had originally been intended to be an eighth 4-4-2T, was added in 1912. Also in 1910, the Hull & Barnsley Railway obtained five domeless 4-4-0s from Kitson & Co. to work its Hull–Sheffield (Midland) through passenger services, which were later

Hull & Barnsley Railway Class J 4-4-0 No. 38 built by Kitson & Co. in December 1910. *Former Ian Allan Archive*

North Staffordshire Railway Class G 4-4-0 No. 86, built by Stoke Works in 1910, at Llandudno Junction.
*F. R. Hebron/Rail Archive Stephenson*

suspended in 1917 and never reinstated. During 1913–14 the Furness Railway put into service four attractive 4-4-0s, built by NBL, which worked expresses between Whitehaven and Carnforth.

NBL also built a total of nine handsome Belpaire 4-4-0s for the Midland & South Western Junction Railway, at intervals between 1904 and 1914, some of which were later fitted with GWR Standard taper boilers. Some years earlier, in 1904, Robert Stephenson & Co. built five 4-4-0s for the Cambrian Railways, one of which was destroyed in the tragic head-on collision at Abermule in 1921. These relatively little-known second eleven 4-4-0s in particular epitomised the sheer highly specialised diversity of the Edwardian locomotive scene.

Furness Railway 4-4-0 No. 131, built by the North British Locomotive Company in 1913, at Barrow.
*Former Ian Allan Archive*

GWR Bulldog class/Bird series 4-4-0 No. 3454 *Skylark*, the penultimate new-build GWR double-framed 4-4-0, which entered traffic in January 1910. *A. W. Croughton/Rail Archive Stephenson*

# GWR 4-4-0s

The Great Western Railway turned to the 4-4-0 in 1894, two years after its abolition of the 7ft gauge, and Swindon produced a total of 369 of these up to 1912.

Of these, forty were of modern Churchward County design with outside cylinders, but not a single example of the remainder had both inside frames and inside cylinders, such as were so numerous everywhere else in the land. Instead, these all sported picturesque double frames, a strangely anachronistic feature that harked back to the now-departed Broad Gauge era. GWR 4-4-0 history and inter-relationships were exceptionally complex. Apart from forty very unusual conversions from unsuccessful 0-4-4 tank engines, and the four Armstrong 7ft 4-4-0 equivalents of the celebrated Dean 4-2-2s that were built in 1894, fundamentally these divided into two main groups, those with either 5ft 8in diameter coupled wheels, the Dukes and Bulldogs, or with 6ft 8½in, the Atbaras. Of the latter, the City series when introduced in 1903 were initially provided with parallel, domeless Belpaire boilers, in which form No. 3440 *City of Truro* was alleged to have broken the 100mph barrier in May 1904.

New double-framed 4-4-0s of both varieties that were built from late 1903 were paradoxically topped off with the sophisticated new standard No. 2, or slightly larger and heavier No. 4 taper boilers, which in due course would

become superheated and adorned with top feed. New construction was rounded off in 1908 with the twenty slightly effete lightweight Flower class engines with No. 2 boilers and 6ft 8½in coupled wheels, which were directly followed in 1909–10 by the rather more purposeful looking fifteen-strong Bird series of 5ft 8in Bulldogs with No. 4 boilers. Whereas the Flowers were withdrawn as early as between 1927 and 1931, as part of a concerted purge by the GWR to eliminate all of its 6ft 8½in 4-4-0s by 1935, the 'Birds' soldiered on intact until 1948.

The forty Churchward County class 4-4-0s, envisaged in January 1901 and fitted with Standard No. 4 taper boilers, in some instances were built virtually alongside their double-framed 4-4-0 counterparts, and one wonders why more Counties were not built instead of these. The experimental fitting of the 3-ton lighter No. 2 boiler to No. 3805 during 1907–09 might have amounted to something of an explorative step in this direction, in the interests of wider route availability. It is known, however, that the County class was notoriously rough riding, which was attributable to its powerful front end design in conjunction with a short (8ft 6in) coupled wheelbase and inherent balancing problems. The Counties had relatively short lives, for in line with the policy mentioned above, all were withdrawn between 1930 and 1933.

GWR County class 4-4-0 No. 3826 *County of Flint*, built in 1906, recorded in its final superheated condition in 1924. *W. V. Wiseman/Rail Archive Stephenson*

Unlike Churchward's ground-breaking 2-6-0 mixed traffic and 2-8-0 heavy goods, which clearly quickly influenced several of his contemporaries, most notably Nigel Gresley on the Great Northern, the Counties did not prompt other companies to build *modern* outside-cylinder 4-4-0s. One possible exception was the Highland Railway, for which two such engines, named *Snaigow* and *Durn*, were designed and built by R & W Hawthorn Leslie & Co. in 1916. These had been envisaged by Frederick Smith, specifically for service on the Far North line between Inverness and Thurso, very shortly before his sudden departure during the previous year.

# 0-6-0

The humble British 'wet steam' 0-6-0 had played a very important, if largely unsung role in underpinning the nation's industrial prosperity during the late 19th and early 20th centuries. Remarkably, the 0-6-0, latterly in many instances superheated, would retain its overall numerical dominance in Britain until as late as 1958 on British Railways, after which it was finally overtaken by the large number of mixed traffic and express passenger 4-6-0s.

Until 1900, the typical 0-6-0 *invariably* had inside cylinders with slide valves that were supplied by a slim (c. 4ft 3in diameter) boiler, and was usually fitted with Stephenson valve gear. Nevertheless, there were some variations on the basic theme. 0-6-0s on the Hull & Barnsley Railway were fitted instead with Allan straight link valve gear, probably as a result of the influence of Kitson & Co., which was its principal locomotive supplier. Meanwhile, from 1889 those on the Lancashire & Yorkshire Railway were provided Joy valve gear, which was already favoured by the LNWR and first employed on its

Cauliflower express goods 0-6-0s introduced in 1880. Joy gear dispensed with eccentrics on the driving axle by working directly from the movement of the connecting rod, albeit at the slight attendant risk of these occasionally breaking at a particularly vulnerable point where the hole was necessarily drilled through it to accommodate an essential motion pin. Some new 0-6-0s built for the HBR in 1900 were fitted with domeless boilers of 5ft diameter, while in 1903 the Midland uprated its long-standing Johnson Class 2 0-6-0 to Class 3, by increasing the boiler diameter on the new engines from 4ft 2in to 4ft 8in and their grate area from 17½ to 21 sq ft. The following year the North Eastern went one better by very substantially enlarging the boiler diameter on its corresponding standard 0-6-0s from 4ft 4in to 5ft 6in in the new P2 class.

Of the coal-hauling railways in South Wales, only the Taff Vale employed 0-6-0 tender engines, although the last of these was built in 1889, after which it favoured instead the 0-6-2 tank engine, the first of which it had received in 1885, and it was a type that also went on to become widely used by other railways in that region. The final 0-6-0s were built for the LSWR as early as 1897, for the LBSCR in 1906, and for the SECR in 1908. The North Staffordshire Railway Class H 0-6-0 (1909) had a close affinity with the SECR Wainwright Class C 0-6-0 (1900), simply because of the transfer of key NSR locomotive personnel from Ashford to Stoke. This was quickly followed by the Belpaire Class H1 during the winter of 1910–11, there being only four engines in each class.

Superheating brought a further boost in haulage capacity to many 0-6-0s, both newly built or when later rebuilt. The second and third locomotives in Britain to be fitted with true superheaters were a pair of 0-6-0s built by the LYR in late 1906. Twenty production engines, likewise

NER Class P2 0-6-0 No. 1687 (5ft 6in boiler), built at Darlington Works in 1904. *Rail Archive Stephenson*

having Schmidt superheaters, were later built at Horwich in 1909. Twenty more with Belpaire fireboxes and sporting three different variants of superheaters between them appeared in 1912. This large group of 0-6-0 locomotives totalled almost five hundred, all of which had been built at Horwich Works spread over a period of nearly thirty years. Yet the first and last of these were almost identical, because a final ten produced during 1917–18 were assembled largely from spare parts and these reverted to John Aspinall's original 1889 design with round-topped firebox. These therefore lacked a superheater, which was also omitted from a contemporary batch of new 0-8-0s.

The coupled wheel diameter on British 0-6-0s usually ranged between 4ft 9in and 5ft 3in, but in 1908 Henry Ivatt on the Great Northern Railway introduced a new class having 5ft 8in coupled wheels, which was later (1912) perpetuated by Nigel Gresley in superheated form. As later determined by the Bridge Stress Committee, the hammer blow values of these two classes came very close behind those of the North Eastern R and R1 class 4-4-0s mentioned earlier. A little earlier, in 1906, the Midland temporarily rebuilt three of its Class 3 0-6-0s with unusually large 6ft diameter coupled wheels for reasons that remain unclear, possibly to handle express fish traffic, or for trial purposes in anticipation of the Deeley front-coupled 0-6-4 passenger tank engines having 5ft 7in diameter coupled wheels that were built the following year.

At 57¾ tons without tender, the heaviest of all the very numerous non-superheated 0-6-0s on various railways was the Peter Drummond 279 class of the Glasgow & South Western Railway. Built in 1913, this was

fitted with Dugald Drummond's steam drier in the smokebox and his system of feed water heating in the tender, which therefore required a feed pump, leading to them becoming known as 'Pumpers'. Despite these accoutrements, David Smith recorded that they burned coal at the prodigious rate of 90lb to 100lb per mile, or sometimes even as much as 120lb, which must have constituted a British record.[5]

A year earlier, the Caledonian Railway had built four more modestly proportioned superheated 0-6-0s, the CR 30 class, which mechanically speaking virtually amounted to being a goods version of the contemporary superheated Dunalastair IV 4-4-0. However, they were painted blue, fitted with the Westinghouse brake, and employed on passenger trains on the Clyde coast. They were quickly followed by a further five very similar engines on which the frames were extended at the front end in order to incorporate a leading pony truck. By contrast, the 34 class 2-6-0s, painted black, were invariably employed on goods trains.

In 1911 the Midland Railway built two new superheated 0-6-0s with piston valves and Belpaire fireboxes. Regular production of these did not begin until 1917. Ultimately no fewer than 772 of the ubiquitous 4Fs were built in total until 1941, including five hundred after 1922 incorporating only minor variations.

Despite having a few years earlier designed an outside-cylinder 0-8-0 to work its Fife–Aberdeen coal traffic, this did not materialise and instead the North British Railway remained content with a modestly proportioned superheated 0-6-0 that was introduced at the end of 1914, of which the last examples would remain in traffic

5 Smith, D. L., *Locomotives of the Glasgow & South Western Railway*, David & Charles, 1976, pp. 117–119.

GNR Ivatt 5ft 8in 0-6-0 No. 11 leaves Potters Bar with a down outer suburban train from King's Cross, c. 1913. This class was sometimes employed on London–Skegness excursions. *Rail Archive Stephenson*

until 1966. Around the same period the Great Eastern produced a superheated 0-6-0 in 1912 that was unusually provided with the same long stroke 20in by 28in inside cylinders as the recent express passenger 4-6-0s. Later, in 1920, this would be further developed to accommodate the 4-6-0's boiler as well.

Of the 1,200-odd 0-6-0s built between 1901 and 1914, Midland Railway No. 3835 was statistically the most significant, although only two such engines were initially built in 1911. Unknowingly at that time these were the precursors of no fewer than 770 further engines that would be built for the MR and LMSR between 1917 and 1941, of which the last remaining examples would cease operation in 1966. *Former Ian Allan Archive*

# 5

# Edwardian Locomotive Design Difficulties and Dilemmas

Modestly proportioned inside-cylinder 4-4-0s and 0-6-0s as built in the 1890s, and sometimes later, with their relatively short boilers having deep fireboxes neatly sunk between the crank and trailing coupled axles, presented no major design problems. Near ideal basic design ratios were still easily achieved, probably almost unconsciously, simply by going on past experience and following locally well-established rules of thumb. Known examples were an approximate allowance of 1 square inch of fire grate area per 4 cubic inches of cylinder volume, and 2½ sq ft of heating surface for each square inch of the area of one piston, when locomotives invariably had two cylinders. As early as 1886 an editorial in *The Engineer* anticipated future problems achieving such simple goals as locomotives became progressively larger.[1] Also, a decade later when it came to enlarging 4-4-0s into 4-4-2s or 4-6-0s, and 0-6-0s into 0-8-0s, having longer boilers with larger fireboxes in association with bigger cylinders, associated design problems indeed arose. Not assisted by the vertical limitations imposed by the restricted British loading gauge, a particular problem could be providing an ashpan of adequate proportions beneath the firebox yet still accommodating a trailing axle. The accumulation of

ash over the necessary hump could also blank off the air supply to the rear portion of the grate, thereby seriously reducing its effective area by perhaps as much as one third. Furthermore, the inevitable close proximity of the hot ashpan to the trailing axleboxes could sometimes prompt these to overheat.

Inside cylinders had ideally suited 4-4-0s and 0-6-0s in particular from the constructional point of view, and not least simply on aesthetic grounds, at that period itself a major consideration with a complete disregard of any questions of accessibility. In 1900, locomotives with outside cylinders were comparatively rare in Britain. Indeed, such were unknown on a number of railways, including most notably the Great Western at that period, but things there would very soon change dramatically in that respect. That said, however, at 31 December 1913, no less than 90 per cent of the 19,000-odd locomotives of the fourteen major railways, out of which roughly three-quarters were accounted for by 4-4-0s, 0-6-0s and 0-6-0Ts, had inside cylinders.

The first moves towards building significantly larger locomotives than hitherto became apparent during the early 1890s, initially regarding goods rather than

Caledonian Railway 4-6-0 No. 49, built at St Rollox Works in 1903. *Former Ian Alian Archive*

[1] *The Engineer*, 13 August 1886, pp. 131–2.

Caledonian Railway 0-8-0 No. 600, built at St Rollox Works in 1901, its coupled wheelbase of 22ft 4in was exceptionally long for a British eight-coupled locomotive. *Former Ian Allan Archive*

passenger locomotives. In 1892 Francis Webb on the LNWR at Crewe built a solitary inside-cylinder 0-8-0 that amounted to an extrapolation of his classic simple 0-6-0 coal engine (construction-wise this had in fact been tacked on at the end of the final batch of these engines). Two years later the first British 4-6-0, with outside cylinders, was built for the Highland Railway, also for goods working, over the heavy gradients between Perth and Inverness. Although in the past sometimes associated with an Indian antecedent designed in Glasgow, a very reasoned case has more recently been made that in reality this owed much to a Manchester-based design for New South Wales.[2] In 1896 the GWR built a one-off inside-cylinder double-framed 4-6-0 with wide firebox, No. 36, known as 'The Crocodile', to work heavy goods trains through the Severn Tunnel with its attendant gradients at each end. Although successful in its objectives, it remained experimental and was withdrawn in 1905. Then followed the enigmatic Krugers. Firstly there was a solitary 4-6-0 with a wide Belpaire firebox built in 1899, which operated for merely five years. Nine very similar 2-6-0s having a pony truck in place of the leading bogie later appeared, eight of them in 1903, which were retired after only *three* years.

Prior to 1896, in British domestic locomotive practice no boiler had been built that exceeded 4ft 6in in diameter, but in that year the Caledonian Railway at St Rollox under John McIntosh built a new 4-4-0 in the classic Dugald Drummond tradition, in which this key dimension was increased to 4ft 9in for the first time to produce the celebrated Dunalastair. The year 1897 witnessed the almost simultaneous appearance on three different railways of the first *four*-cylinder locomotives, on 4-4-0s on both the Glasgow & South Western in April, and the London & North Western in June, which were closely followed in August by an uncoupled 4-2-2-0 equivalent on the London & South Western. In May 1898 the first British 4-4-2 tender engine appeared on the Great Northern Railway, to be followed in June 1899 by the first British passenger 4-6-0 on the North Eastern Railway, both of which had outside cylinders.

If not quite as large as the North British, the Caledonian nevertheless regarded itself as Scotland's leading railway, and under John McIntosh it quickly followed the new large locomotive trend. In June 1900 an inside-cylinder 0-8-0 heavy goods engine was outlined, the first two of which were completed just over a year later. Then, having in September 1901 briefly considered an inside-cylinder express passenger 4-4-2, this was later discarded in favour of a somewhat heavier 4-6-0 proposed in April 1902. This duly appeared in March 1903 as CR No. 49. The 0-8-0 represented a considerable advance on the standard 812 class 0-6-0 introduced only two years earlier in 1899, which it is understood that McIntosh, as a 'running man', always considered to be his favourite locomotive. The boiler for the 0-8-0 was very closely based on that of the so-called Dunalastair III 4-4-0, whose boiler barrel was simply considerably extended by almost 40 per cent. The boiler for the 4-6-0, on the other hand, was clearly newly designed and actually influenced that for the *later* and smaller Dunalastair IV 4-4-0, which appeared in May 1904.

[2] Rutherford, M., David Jones of the Highland Railway and the writers: forerunners of the 'Big Goods' 4-6-0, *Back Track*, February 2007, pp. 99–108.

### Table 11 Caledonian Railway 600 class 0-8-0 and 49 class 4-6-0 dimensions compared to those of CR 0-6-0 and 4-4-0 designs

| CR class<br>Date | 812<br>0-6-0<br>1899 | 600<br>0-8-0<br>1901 | Change | 140<br>4-4-0<br>1904 | 49<br>4-6-0<br>1903 | Change |
|---|---|---|---|---|---|---|
| Cylinders (2) inside, in x in | 18½ x 26 | 21 x 26 | | 19 x 26 | 21 x 26 | |
| Cylinder vol. ft³ | 8.09 | 10.42 | +28% | 8.53 | 10.42 | +22% |
| Boiler pressure, lb | 160 | 175 | +9% | 180 | 200 | +11% |
| Evaporative HS, ft² | 1403 | 2108 | +50% | 1615 | 2323 | +44% |
| Grate area | 20.6 | 23.0 | +12% | 21.0 | 26.0 | +24% |
| Boiler free gas area | 3.44 | 3.57 | +4% | 3.67 | 3.68 | 0% |
| as % of grate area | 16.7 | 15.5 | | 17.5 | 14.2 | |
| Boiler diameter, ft in | 4 9 | 4 9 | 0% | 5 0 | 5 0 | 0% |
| Tube length, ft in | 10 7 | 15 7½ | +48% | 11 6 | 17 3 | +50% |
| Engine weight, tons | 45.7 | 60.6 | +33% | 56.5 | 73.0 | +29% |

On all four designs, 1¾in diameter fire tubes were employed, regardless of their length (on the 49 and 140 classes a few tubes of a slightly larger diameter were fitted in the bottom of the boiler, where they were more prone to blockage with ash). The optimal length to bore ratio, *as was later established*, was about 100: 1, giving an ideal length for a 1¾in diameter tube of only about 12ft. It is therefore difficult to imagine that the 4-6-0 boiler as originally designed would have been a very satisfactory steamer, having such long tubes of this diameter. Presumably, in fairly prompt acknowledgement of this shortcoming, a third boiler was very quickly designed and built in 1904 for use on the two 4-6-0s, in which the length between tubeplates was now substantially reduced to 15ft 8in, while the now uniform tube diameter was increased to 2in. According to official figures, this development also reduced the original engine weight by a surprising 3 tons. At the same time the cylinders on both engines were reduced in diameter from 21 to 20in, and their boiler pressure likewise from 200 to 175lb, which when combined resulted in a theoretical power reduction of 20 per cent.

In the subsequent five CR 903 (Cardean) 4-6-0s built in 1906, the original boiler pressure of 200lb was restored but in association with 20in diameter cylinders. The boiler diameter was also increased from 5ft to 5ft 3½in, while the tube length was increased again to 16ft 8in, for which the 2in diameter tubes were just about appropriate. Significantly also, the ashpan was now provided with both front *and* back dampers. The *driving* wheel centres were dished outwards to permit longer journals.

However, both the 0-8-0 and original 4-6-0 designs had anticipated an increase in power output of the order of 30 per cent in return for only a negligible increase in boiler free gas area through the fire tubes. The latter dimension could be a limiting factor regarding steaming capacity, and could incur counterproductive excessive back pressure in the cylinders, which thereby reduced the potential power output. The concept of free gas area and

its significance was at that time scarcely recognised. (Most unusually 'flue area' was quoted on several general arrangement drawings and some diagrams for Peter Drummond locomotives both actual and proposed on the Highland and Glasgow & South Western railways from as early as 1899 onwards.) In the 1920s, Richard Wagner in Germany established that the optimal value, at least in association with simple exhaust systems, was equivalent to 15 per cent of the grate area. In this respect, none of the four boilers shown above was too badly out of kilter. This important factor was addressed only extremely briefly in but a single short paragraph, which also quoted misleadingly high 'typical' values, by E. A. Phillipson in his book *Steam Locomotive Design: Data and Formulae*, published in 1936.

An indication of the wide actual range was provided by two British non-superheated 4-4-0s, in which this factor varied from as low as only 9.3 per cent in the LSWR Dugald Drummond T9 4-4-0 (1899) with a 4ft 5in diameter boiler and 24 sq ft grate area, and to no obvious detriment, to as high as almost 20 per cent in the LNWR Precursor (1905), whose corresponding dimensions were 5ft 2in and 22.4 sq ft. Somewhat later, the superheated Derby Class 2 simple and Class 4 Compound 4-4-0s shared precisely the same tube and flue arrangement but in association with respective grate areas of only 21.2 and 28.4 sq ft, which resulted in corresponding gas area/grate area proportions of 16.4 and 12.2 per cent. In this regard it is interesting to note that whereas the Somerset & Dorset 2-8-0s built at Derby in 1914 also utilised the Compound boiler, the second batch of these when built in 1925 was provided with a new boiler, whose diameter was increased from 4ft 8in to 5ft 3in. This retained the same grate area, but related to it the free gas area was now elevated to a notably more generous 17.2 per cent. These later 2-8-0s actually copied a very similar but abortive proposal that had already been made by the Midland Railway in 1919 for its own use. However, to have extended the use of the larger boiler to the

Great Central Railway Director class 4-4-0 (1913 series) No. 431 *Edwin A. Beazley. Former Ian Allan Archive*

Compound 4-4-0s would have raised their axle load to a probably unacceptable 21 tons or so. Having said this, in recent years a previously unknown outline scheme dated 1925 for a proposed LMS 'Super Compound' has come to light. With an anticipated coupled axle load of 22½ tons, and a grate area of 31½ sq ft, this would have been the largest British 4-4-0 ever built.

On a weight basis, at 75 tons the Sir Sam Fay superheated 4-6-0s designed under John Robinson on the Great Central Railway in 1912, ranked as the largest British inside-cylinder locomotives, in which the cylinders were of 21½in diameter. It has sometimes been claimed that the more successful Director class 4-4-0 (with 20in cylinders) that quickly followed nine months later, had been hastily designed upon the almost immediate

recognition that the big 4-6-0 was a failure. In fact, both had been designed almost simultaneously, the 4-4-0 being intended to supersede the famed 4-4-2s. However, it is not hard to see why the 4-6-0s were disappointing. Despite a substantial 15½ per cent increase in cylinder volume, the grate area remained unchanged at 26 sq ft. This latter dimension had first been reached on the GCR by its *commercially* designed first 4-4-2 in 1903, which achieved an excellent match of its cylinder capacity (via a 26in piston stroke with short travel slide valves) to a well-proportioned parallel 5ft diameter 180lb pressure boiler. In this respect it was comparable with the exactly contemporary yet distinctly more sophisticated GWR Churchward Saint 4-6-0, with its 30in stroke cylinders served by large-diameter, long-travel piston valves and a

GCR 4-6-0 No. 423 *Sir Sam Fay* at Guide Bridge, probably when new in 1913. *P. F. Cooke/Rail Archive Stephenson*

225lb pressure taper boiler (5ft 6in maximum diameter). Both boilers had a similar tube length of 15ft, and likewise contained 2in diameter tubes. When compared with the Director 4-4-0, the boiler on the new GCR 4-6-0 was very substantially increased in its length between tubeplates by 5ft from 12ft 7in to a truly excessive 17ft 7½in (the tube diameter was at least correspondingly increased from 1⅞ to 2¼in). This combination resulted in a disproportionately larger but less effective increase in evaporative heating surface in relation to the grate area, which was also associated with an enlargement in boiler diameter from 5ft 3in to 5ft 6in.

Over the next few years the boiler tube and flue arrangement on these large 4-6-0s was revised twice, in attempts to boost the degree of superheat. Meanwhile, after several years their unusually large 21½in diameter cylinders were reduced in bore to 20in. The 4-6-0s are said to have suffered from heating problems in the driving axleboxes owing to their restricted dimensions. In both the 4-4-0s and 4-6-0s the cylinders were pitched at 2ft 0½in centres and the axle load of the driving axle was very similar at 19.9 and 19.5 tons respectively. However, the journals on the 4-6-0 were actually more generous at 9in diameter by 9in long, compared with only 8in by 9in on the four-coupled engines. (For comparison, on the McIntosh 49 class 4-6-0s these had been 9½in by 9½in.) Assuming that the weights of both GC driving wheelsets complete with crank axles were 5 tons, then the *static* vertically acting bearing pressure on the driving journals of the Robinson 4-6-0 at 200 lb/sq in would actually have been about 15 per cent *lighter* than that on the 4-4-0. This value was well down the scale within the range of 180 to 250lb/sq in advocated by E. A. Phillipson in his book. However, when in motion the journals on the 4-6-0 would have been subjected to an almost 8 per cent greater near horizontal maximum piston thrust per sq in from the 21½in diameter cylinders, when compared with those of the 4-4-0, which presumably must have been the real problem, and which later led to the reduction in the cylinder bore.

Almost certainly drafted in early 1912 as an alternative to the Sir Sam Fay 4-6-0 design was a superheated *narrow firebox* 4-6-2 with 22in diameter *outside* cylinders that was also outlined at Gorton.[3] This would have incorporated a similar but 2¼ft shorter 5ft 6in diameter boiler when compared with that of the 4-6-0, but with a ½ft longer firebox that afforded 27.7 sq ft of grate area, which was to be centred above the trailing coupled axle. Although weighing an estimated extra 4½ tons compared with the 4-6-0, the maximum axle load of the 4-6-2 would have been only 18 tons.

Although strictly speaking falling outside the scope of this book, despite a wartime embargo on express passenger locomotive building, sanction was somehow obtained to build a prototype *four*-cylinder version of *Sir Sam Fay* at Gorton in 1917. This would have divided cylinder drive, producing individual 45 per cent lighter piston thrusts and having the slightly smaller 'Director' size 8in by 9in leading driving journals. Its hammer blow would be only 2.3 tons, as against 5.5 tons at 74mph, although, unlike on certain other railways, matters of weight do not seem to have unduly concerned the GC civil engineer. Named *Lord Farringdon* after the newly ennobled GCR chairman, and designated Class 9P, the new 4-6-0's four 16in by 26in cylinders nevertheless increased the *total* cylinder volume by a further 10 per cent.

As if this were not enough, the general arrangement drawing for a mixed traffic version of *Lord Farringdon*, with 5ft 8in in place of 6ft 9in coupled wheels, that was later introduced in 1921 (GCR Class 9Q) specified a boiler pressure of 200lb, compared with the previous standard 180lb. All things combined, this implied an expectation for an increase of 40 per cent in power output for no increase whatever in grate area, when compared with the well-proportioned Director 4-4-0.[4] In the event, the 9Q boilers actually operated at the standard 180lb. These nevertheless popular 4-6-0s were quickly dubbed Miners' Friends and Black Pigs on account of their healthy appetite for coal, which they unsurprisingly shared with their earlier 9P passenger counterparts. This can readily be attributed to the inadequate 26 sq ft grate area promoting excessively high firing rates per square foot, thereby resulting in correspondingly reduced boiler thermal efficiency. This was further aided and abetted by the generous boiler free gas area, which amounted to almost 20 per cent of the limited grate area, and which would therefore have been conducive to the ejection of unburned coal through the chimney.

These shortcomings were further compounded by poor cylinder performance resulting from the standard short 1in lap/4½in maximum travel of the piston valves with their closely pitched valve heads, in conjunction with long and tortuous S- shaped steam passages between the cylinders and steam chest above. (S could equally stand for strangulated.) Seemingly in belated recognition of this, somewhat improved cylinders were actually fitted to the last few 9Qs that were completed under the LNER. Although these were also retro-fitted to some earlier 9Q engines, even until as late as 1946, curiously they were never applied to any of the six express passenger 9Ps, of which five more were built in 1920. An interesting feature of both GCR four-cylinder 4-6-0 classes was that the

---

[3] Jackson, D., *J. G. Robinson – A Lifetime's Work*, The Oakwood Press, 1995, p. 219, proposed GCR 4-6-2 diagram.

[4] As late as 1945 the GWR introduced the Hawksworth County class 4-6-0, which retained the standard 18½in by 30in cylinder dimensions of the preceding Saint and Hall classes, while the boiler pressure was increased from 225 to 280lb (by 24 per cent) the grate area correspondingly only rose from 27.1 to 28.4sq ft (by 5 per cent). These engines did not achieve the high reputation of their predecessors. Their boiler pressure was later reduced to 250lb, which together with revised tubing arrangements and later re-draughting with double blastpipes and chimneys, they were finally much improved by the late 1950s, shortly before their early demise.

One of the five production Great Central Railway four-cylinder 4-6-0 express passenger locomotives, built at Gorton Works in 1920, No. 1167, while briefly named *Lloyd George* until 1923, before that politician fell from grace, passes Rushcliffe Halt, south of Nottingham, with a London express. *Former Ian Allan Archive*

adjacent piston valves on each side of the engine were worked in unison by one set of valve gear, in a manner very similar to that on the NER 4CC class 4-4-2s yet to be described, but in a simple expansion context. The link was John Smith, who was the works manager at Gorton from 1906 until his retirement in 1932, and who was the son of Walter M. Smith, designer of the 4CCs.

Simple arithmetic suggests that pro rata the 8½ft long fireboxes on these three large Great Central 4-6-0 classes should have been made at least 1ft longer in order to yield a grate area of around 30 sq ft.[5] Such had been one redeeming feature of the contemporary LNWR Claughton four-cylinder 4-6-0s, on which it was 30½ sq ft. However, there is a possible clue as to the reason for Robinson's apparent reluctance to increase the grate area. Back in 1904, when he took part in a discussion of French compound locomotives, he recalled experiencing a recent impressive run in France behind a Paris–Orleans compound 4-4-2. This had 33½ sq ft of grate area, from which he had calculated that the firebox must have been 'enormously

long'.[6] Its outside length was in fact only 2in shy of 11ft, i.e. almost 2½ft longer than the established standard on the Great Central engines.

Fifteen years later there was surely a certain irony when William Rowland, the chief locomotive draughtsman at Gorton Works, who would have overseen the preparation of all these 4-6-0 designs, delivered an erudite paper to the Institution of Locomotive Engineers in November 1919 on the subject of boiler performance and efficiency. This contained numerous mathematical formulae and Rowland gave as a worked example the Sir Sam Fay boiler, with its *final* tubing arrangement with twenty-eight-element superheater, operating at the high, although by no means exceptional firing rate of 129lb per sq ft of fire grate per hour. The very precision of this quoted figure, which corresponded to an hourly *coal* rate of 3,354lb, rather suggests that this level might have been attained during tests.[7] There is photographic evidence that indicator trials at the very least were undertaken on No. 423 *Sir Sam Fay* when it was new.

[5] In 1955 W. A. Tuplin published the first of his many analyses of how various British locomotive classes might have been designed to better effect. In 'The "Sir Sam Fay" Class: An Appraisal', *The Railway Magazine*, April 1955, pp. 234–8. Tuplin advocated a grate area of 31 sq ft for this class.

[6] Sauvage, E., Compound Locomotives in France, *Proceedings of the Institution of Mechanical Engineers*, 1904, pp. 327–467.

[7] Rowland, W., An Approximate Method of Estimating Superheat and Boiler Output and Evaporative Efficiency, *The Journal of the Institution of Locomotive Engineers*, No. 41, October–December 1919, pp. 459–69.

London & North Western Railway Precursor class 4-4-0 No. 737 *Viscount* awaits departure from Euston.
*Former Ian Allan Archive*

For its part, comparisons had earlier been made by the London & North Western Railway of both the design and performance of its inside-cylinder Precursor 4-4-0 and Experiment 4-6-0, designed under Francis Webb's immediate successor, George Whale. These had been introduced in 1904 and 1905 respectively, when indicator trials were almost immediately conducted with each prototype. An average indicated horsepower (IHP) in the cylinders of 1,002, with a maximum IHP of 1,197 was recorded by the 4-4-0, whereas the corresponding values for the larger 4-6-0 were both approximately 15 per cent lower at 880 and 994 IHP respectively.

The greater horsepower power developed by the slightly smaller 4-4-0, with 6ft 9in diameter coupled wheels, was not entirely surprising given its less inhibited boiler design, i.e. the shorter tubes and also the shorter but 14in deeper firebox sunk between the coupled axles, which also made for a simpler ashpan layout. As originally designed the free gas area approached 20 per cent of the 22.4 sq ft grate area. On the 4-6-0, with 6ft 3in coupled wheels, partly by virtue of the slightly shorter connecting rods, the tube length was increased by only 7¾in. The firebox was 10in longer to give a horizontal grate having an area of 25sq ft, but this was much shallower, being set directly above the centre and trailing coupled axles. The ashpan was therefore now more restricted, but on the credit side slightly fewer fire tubes and the larger grate area combined to reduce the free gas area to grate area proportion to a more ideal 16.4 per cent.

Crewe later engaged Dr F. J. Brislee of the University of Liverpool to undertake[8] smokebox gas analysis which showed the 4-6-0 boiler to be superior as regards combustion efficiency on account of the necessarily thinner fire, which was much closer to the firebrick arch (and which thereby required more skilled firing). Dr Brislee discovered that the air supply to the much thicker fire in

the deeper 4-4-0 firebox to be a limiting factor, and suggested that th s be augmented. Crewe drawing office thereupon produced a scheme for a proposed belt-driven air blower that would be mounted on the front of the tender linked by a flexible duct to the ashpan. There is no evidence that this proposal was ever put into effect.

When Dr Brislee read his paper to the Institute of Mechanical Engineers in March 1908, during the ensuing discussion Charles Bowen Cooke, standing in for George Whale, stated that during the previous five months the coal consumption and timekeeping of a representative Precursor 4-4-0 No. 276 *Doric,* and an Experiment 4-6-0 No. 1987 *Glendover,* had been monitored while each ran some 34,000 miles. Respective coal consumption had been 57.5 and 52.25lb per train mile, giving a 10 per cent fuel economy in favour of the Experiment. As regards timekeeping, the Precursor had on average gained seventy-nine seconds on the schedule compared with sixty-nine seconds by the smaller-wheeled Experiment, a fellow engine from which class would later be credited with running at 9Cmph downhill at Shap.

In 1911, the Prince of Wales superheated development of the Experiment class was introduced, which became the largest British *superheated* express passenger locomotive design to be equipped with Joy valve gear. By 1922, when 245 of these engines had been built, this feature was an anachronism that had resulted in the fracture of a connecting rod on one locomotive, with fatal consequences for both engine crew. Former Crewe LNWR locomotive practice had little influence on subsequent LMSR locomotive policy but, its inside cylinders notwithstanding, surprisingly there was a serious proposal as late as 1931 to build a new modernised version of the Prince, to be fitted with Caprotti poppet valve gear and provided with mechanical lubrication.[9] This proposal was very quickly abandoned after William Stanier took over as

[8] Brislee, F. J., Combustion Processes in English Locomotive Fire-boxes, *Proceedings of the Institution of Mechanical Engineers*, 1908, pp. 237–268.
[9] Cox, E. S., *Locomotive Panorama*, Vol. 1, Ian Allan, 1965, diagram of 4-6-0 on p. 81.

LNWR Experiment class 4-6-0 No. 507 *Sarmatian* at Euston, c. 1906. *Rail Archive Stephenson*

CME the following year, and was replaced by his own very up to date Class 5 6ft mixed traffic 4-6-0, which was nevertheless initially referred to in early documentation as either *Converted* or *Improved* Prince of Wales class![10]

Also back in 1908, on the North Eastern Railway a brief reversion from 4-4-2s to large 4-4-0s had produced very disappointing results. The R1 class was probably built as a substitute, having roughly equivalent power, for the ten additional four-cylinder compound 4-4-2s that had been authorised only three months earlier. These had not been proceeded with, allegedly because of patent royalty issues with the late W. M. Smith's estate. The R1s were originally intended to have the same 7ft 1¼in coupled wheel diameter as the 4CCs, and initially they did carry the same very high boiler pressure of 225lb. After the emergence of the first three locomotives in late 1908, there was a significant delay of several months before delivery of the remaining seven resumed from Darlington Works. George Heppell's memoirs appear to imply that the latter may actually have been completed on schedule, only to have to be dismantled again to make good the poor workmanship that had become apparent in the first three engines. He made no specific allusion, however, to something recalled forty years later by Edward Thompson when he was interviewed by Brian Reed, to the effect that the coupling rods for the first engine, No. 1238, had been made ¼in in excess between crank pin centres, possibly with malicious intent. Whatever happened, the upshot was the sudden departure of the Darlington works manager, Ramsay Kendall, a former colleague of Vincent

Raven, who by 1908 was Wilson Worsdell's deputy.

Heppell did, however, recall being called into Raven's office, where he was heavily criticised over the R1 cylinder design. The axes of the cylinders and piston valves mounted above them diverged markedly, which resulted from the adoption of direct drive outside admission Stephenson valve gear. Heppell characteristically stood his ground and unashamedly put his superior firmly in his place.[11] Nevertheless, Raven did have a point, for this feature made achieving satisfactory valve events at both ends of the cylinders well-nigh impossible. Heppell conveniently overlooked the various alterations that were quickly made to the valve settings on the first R1s when they were newly in service.[12] These attempts to improve them in fact evidently only made matters worse. Many years later in 1942, when he was newly CME of the LNER, Edward Thompson proposed rebuilding the R1s with new cylinders and long travel valve gear. However, withdrawal of the class began at the end of that year.

It is interesting to note that a very similar arrangement was soon also adopted on the original 1913 series of the Great Central Director 4-4-0s. However, in the post-war Improved Directors the main improvement, other than the visible provision of an attractive side window cab, was simply a reversion to the more usual inside admission indirect Stephenson valve gear, operating through rocking levers. Such had been employed as a matter of course in the heavy suburban 4-6-2Ts and Sir Sam Fay 4-6-0s that design-wise had immediately preceded the 1913 Directors.

[10] Cook, A. F., *LMS Locomotive Design and Construction*, The Railway Correspondence and Travel Society, 1990, p. 71.

[11] Heppell, G., *North Eastern Locomotives: A Draughtsman's Life*, p. 17.

[12] David Gray (Darlington draughtsman) notebook in the NRM archives.

NER Class R1 4-4-0 No. 1242 at Neville Hill shed, Leeds, when new in 1909. This design perpetuated the unusual coupled wheel diameter of 6ft 10in, which was peculiar to the NER, of the preceding Class R 4-4-0s and Class V 4-4-2s, and which was later also employed on the three-cylinder Class Z 4-4-2s. *Rail Archive Stephenson*

The NER R1 boilers were also poor steamers, although at 27 sq ft these were certainly not deficient in regard to grate area, at least at the beginning of a run before ash began to accumulate beneath the very shallow rear portion of the ashpan. The R1 boilers were simply truncated versions, with a necessarily shallower firebox, of the Class V 4-4-2 boiler. Therein lay a very simple mistake: at only 11¼ft between tubeplates these retained the same arrangement of 2in diameter tubes, which had been more appropriate to the 5ft longer 4-4-2 boiler, which was 16¼ft between tubeplates. On the preceding smaller highly but successful Class R 4-4-0 introduced in 1899, the tubes had been an ideal 1¾in diameter. Later research in the USA on the Pennsylvania Railroad, and in Germany by R. P. Wagner established that the optimal length to bore ratio for fire tubes was 100 to 1.

In January 1910 a scheme was outlined for a proposed three-cylinder version of the R1, and by early August it was agreed to obtain twelve of these engines (non-superheated) from contractors. However, the tube problem was later seemingly quite unwittingly acknowledged in a report dated November 1910, which concluded:

'The R1 engines are not as economical as the Class V engines when working our East Coast trains … The Class R1 boiler is incapable of evaporating the same quantity of water as that of the Class V, and is overtaxed in evaporating the amount required to work any but our lightest trains … to such an extent as to render it wasteful in coal … future locomotives for heavy express passenger service must be of the Atlantic type, *on account of the length of boiler required*.'

Confusingly, quotations had *already* been sought the previous month for twenty three-cylinder versions of the Class V 4-4-2, which was initially designated V2 at the design stage, but which on its appearance was decreed to be Class Z. The order went to NBL, with half of the engines to be fitted with the Schmidt superheater, while the remaining ten were to be non-superheated for comparative purposes. The superheater engines quickly demonstrated a fuel economy of 20 per cent, and so the non-superheated engines were all converted to conform to them three years later, between May 1914 and April 1915.

Early in 1910, when his first, and indeed Scotland's first, superheated locomotive, CR 4-4-0 No. 139, was taking shape in St Rollox Works, John McIntosh made a rare venture into print in *Cassier's Magazine* for March 1910, with an illustrated article entitled 'British Express Locomotives'. It was clearly too early for him to make any informed judgements concerning the advantages or otherwise of superheating. However, on the question of compounding, in which British interest had patently very recently evaporated, he opined 'the four-cylinder Compound has worked well abroad, although it seems likely that some of its best work is done when running as a four-cylinder simple. Efficiency and reliability are more to a Locomotive Superintendent than economy in fuel'. Back in 1900 McIntosh himself had initiated designs for a four-cylinder compound 4-4-0, with piston valves for all four cylinders, unlike the Webb LNWR 4-4-0s, which had piston valves on the high-pressure cylinders, and slide valves for the low-pressure. After quite a number of drawings had already been made, the CR proposal was suddenly abandoned. In 1905 McIntosh had also briefly looked at a possible de Glehn-style compound 4-4-2, but five simple expansion 4-4-2s were actually authorised.

However, five years on, elsewhere in his article McIntosh also declared that 'the Atlantic type has not been a conspicuous success anywhere', a statement with which his three counterparts on the East Coast partnership would no doubt have disagreed. On the question of inside *versus* outside cylinders he declared 'outside cylinders had apparently some decided advantages, but the greater steadiness of running (of the inside cylinder engine) is due to the closeness of the centres of the cylinders, probably balances all its other defects'.

Less than a year earlier, and possibly after he had already drafted his article, during the evening of 2 April 1909, the Caledonian Railway's flagship express locomotive, 4-6-0 No. 903 *Cardean*, had suffered a very major defect indeed. Having just passed through Crawford with the prestigious Down Corridor express its crank axle fractured at its junction with the left-hand leading driving wheel, which thereupon sheared off and left the engine, which then derailed and was also parted from its tender, fortunately without any serious consequences. The engine had logged an estimated 145,389 miles since new almost three years earlier, and the crank axle had been guaranteed by its manufacturers in Newcastle for 200,000 miles. It was no doubt significant that somewhat later, in February 1911, St Rollox drawing office prepared an outline scheme for a new superheated *outside*-cylinder 6ft 6in 4-6-0, for which it also designed the cylinders in some detail, an indication of serious intent to build.

In the event, although actually shelved, the new 4-6-0 proposal eventually saw the light of day five years later under McIntosh's successor, William Pickersgill, late of the Great North of Scotland Railway. Immediately after Pickersgill had taken over in June 1914, the 4-6-0 was revised with 6ft 1in in place of 6ft 6in diameter coupled wheels and a smaller six-wheel tender. The result was the 60 class of legendary lethargy, of which No. 60 itself was completed in late 1916. Proposals for smaller superheated outside-cylinder 2-6-0 and 4-6-0 mixed traffic locomotives with 5ft 9in diameter coupled wheels had also been made under McIntosh back in 1912. A superheated version of the existing 908 class 5ft 9in inside-cylinder 4-6-0 (1906) was actually adopted instead, which became the 179 class, while proposals to also superheat the older 908 class did not materialise. However, the smaller outside-cylinder 4-6-0 proposal also later appeared, at least in part, under the Pickersgill regime in 1917, but only after it had been modified to become a 4-6-2 tank engine.

*Cardean* had appeared in May 1906, during a particularly interesting few months for the contemporary observer. Only three months later, *Aberdonian*, the first of the equally imposing North British Railway outside-cylinder 4-4-2s made its debut. Meanwhile, in April 1906 a one-off four-cylinder 4-4-2, later named *North Star*, had been completed by the Great Western Railway at Swindon. This had been designed to equal in performance with simple expansion that of the three compound 4-4-2s that the GWR had recently imported from France for evaluation purposes. Simultaneously, the Gateshead Works of the North Eastern Railway outshopped the first of two nameless four-cylinder compound 4-4-2s, which one suspects could well also have been designed as another British repost to the French engines.

John McIntosh himself was neither a locomotive designer nor even an engineer as such, but he had been a former locomotive driver and was possessed of an intimate practical knowledge of locomotive operation. He had lost his right arm in an accident on the footplate, but despite this major personal setback he nevertheless rose through the ranks to be appointed chief running superintendent on the Caledonian Railway in 1891, and its locomotive superintendent four years later. In 1978, no less than sixty years after McIntosh's death, the author was truly astonished to encounter someone who actually retained fond personal memories of him! Graeme Miller had entered St Rollox as a premium apprentice in September 1911, and he provided the following evocative recollection of his former mentor:

> 'His handsome appearance befitted the position he held; tall, broad shouldered and complete with his usual flat-topped hard hat. His right shoulder tilted slightly upwards as from there began his black-gloved artificial arm, a reminder of his early career. Regularly he arrived at the workshops by street car (tram) around 8.30am and, meeting his works manager, William Urie, toured the various departments, before returning to his office. This practice kept each foreman on his toes, as from 6am he would be checking all work in progress in anticipation of any questions that might be asked. Mr McIntosh had started his career at footplate level and had a complete understanding of practical needs, which seemed to compensate for any lack of technical skill. For the latter he had complete faith in his staff and by virtue of his pleasant personality this trust was amply appreciated.'[13]

McIntosh's protégé, Graeme Miller, finally became the chief locomotive draughtsman at St Rollox Works in 1939, which he remained until his retirement in 1959. His railway career had therefore spanned nearly fifty years from the late Edwardian era on the Caledonian Railway, then at its zenith, to the very beginning of the end of the steam era on British Railways, which he would go on to outlive by several years.

[13] G. R. M. Miller to author, May 1978.

# 6
# 4-4-2 versus 4-6-0

## Pre-1901, the first British 4-4-2 and 4-6-0 express passenger locomotives

Express passenger train weights increased significantly during the 1890s, not so much due to an increase in passenger payload, but because of the transition from non-corridor six-wheeled stock to corridor bogie stock, the increasing incorporation of dining cars and sleeping cars, and increases in average running speed. Thus the Great Northern Railway calculated that between 1890 and 1896 alone the average weight of its express

trains between London and York had increased from 170 to 237 tons, and their average speed from 51¼ to 55¾mph, thereby requiring an increase of at least 50 per cent in power output from a single locomotive, if resort to double-heading was to be avoided.

Although the last British 4-2-2 locomotives (with inside cylinders) were actually built by the GNR as late as 1901, the final batch of Patrick Stirling's (outside-cylinder) 8ft 4-2-2s,

Beyer, Peacock & Co.'s as published dual official portraits of the alternative 4-4-2 and 4-6-0 express passenger locomotives that it both designed and built for the Great Central Railway in 1903.

Great Northern Railway 4-4-2 No. 990 as built in May 1898. The name *Henry Oakley*, who had been general manager of the GNR between 1870 and 1898, was later applied in June 1900. *Former Ian Allan Archive*

that had it built only six years earlier, from the start had barely been equal to their task. These had an *official* axle load and therefore adhesive weight of but 19½ tons (although in reality this was probably exceeded). Stirling died while still in office in late 1895, and was succeeded by Henry Ivatt, who very quickly produced a 4-4-0 in 1896, whose axle load amounted to only 14½ tons, while its adhesive weight was 28 tons. These figures would increase in successive but really rather feeble GNR 4-4-0 designs, reaching 18 tons and 35½ tons respectively in his final superheated series built in 1911.

The 4-4-0 never featured as prominently on the Great Northern Railway as it did on several of the other large railway companies. Legend has it that shortly before leaving his previous post in Ireland as locomotive superintendent of the Great Southern & Western Railway, Ivatt had already prepared a diagram in Dublin of a proposed 4-4-2, or 'Atlantic', for his new employer in England. When completed at Doncaster Works in May 1898, Great Northern Railway 4-4-2 No. 990 (now preserved) was the first British tender locomotive to run on *five* axles, and at 58 tons was also but briefly the heaviest so far built, yet it had an axle load of no more than 16 tons.

No further 4-4-2s were built by the GNR until 1900 when it completed a further ten. In the meantime, Ivatt's close friend and predecessor at Inchicore, in Dublin, John Aspinall, now serving the Lancashire & Yorkshire Railway, without building a prototype, first boldly completed a batch of no fewer than twenty 4-4-2s for the LYR at Horwich Works during 1899. These were followed by

another twenty in 1902. This lofty design, with its high-pitched boiler above 7ft 3in diameter coupled wheels, was very different from No. 990 on the GNR, also in that it had inside rather than outside cylinders, and sported a Belpaire firebox. Although *looking* heavier than GNR No. 990, LYR No. 1400 actually weighed only ¾ ton more, but did have an extra 4 tons usefully resting upon its coupled axles available for adhesive purposes.

In June 1899, the GNR's northern English neighbour in the East Coast alliance, the North Eastern Railway, produced its first 4-6-0, No. 2001 (Class S), which was also the first British *six*-coupled passenger engine. Not entirely dissimilar in appearance to the then still solitary GNR 4-4-2, especially around the cylinders, this had the benefit of 46¼ tons as against only 31 tons adhesive weight, and coupled wheels of only 6ft 1¼ in diameter as against 6ft 8in. The NER had to contend with the up 1 in 96 gradient at Cockburnspath while exercising its running powers over the North British Railway north of Berwick, when its locomotives were working up East Coast expresses through from Edinburgh. This right had only very recently been legally contested, unsuccessfully, by the NBR in the aftermath of the 1895 so-called 'Railway Races to the North' (which had not been to Edinburgh but far beyond to Aberdeen). On the Great Northern the only serious gradient encountered between London and York was 1 in 176 north bound below Stoke Summit, south of Grantham.

Where the NER 4-6-0 fell short of its GNR 4-4-2 counterpart, particularly having regard to its own larger cylinders and higher (200lb) boiler pressure, was its

North Eastern Railway Class S 4-6-0 No. 2003 built in September 1899, but shown as rebuilt in June 1901 with longer cab with two windows, in place of only a single side window, together with receiving a longer standard tender.
*Rail Archive Stephenson*

distinctly smaller and shallower firebox. The first three examples of Class S were initially turned out with short cabs and short tenders so that they could be accommodated on existing turntables, but later engines from this initial batch of ten had double side window cabs and longer tenders, and of which the last two also had piston valves in place of slide valves. However, their reign hauling the prestigious East Coast expresses was brief, initially because they were very rapidly superseded in little more than twelve months by a larger-wheeled version, the S1 with 6ft 8¼in coupled wheels and piston valves, of which only five were built. The S1 likewise would also feature only briefly on top link workings between York and Edinburgh. A further thirty cosmetically slightly modified Class S 4-6-0s were later built by the NER between 1906 and 1909 for fast goods work, of which the later ones were therefore painted black.

# After 1900

In 1902 the GNR built two 'experimental' 4-4-2s at Doncaster, No. 271 in July with four cylinders, and No. 251 (now preserved) in December with a much larger boiler having a wide firebox. No. 271 had been ordered back in late 1899, and as built an adjacent pair of inside and outside cylinders shared one piston valve spindle carrying *four* 6½in diameter piston valve heads worked by Stephenson valve gear located between the frames. This arrangement was remarkably similar to that experimentally applied to the

rebuilt LNWR Experiment 4-6-0 No. 1361 P*rospero* in 1915, according to C. F. Dendy Marshall's patent of 1912. Piston stroke was only 20in to correspond with the 10in crank pin throw of the standard 6ft 8in coupled wheel castings (although all other GN 4-4-2s had 24in stroke). The early piston valves were not successful and during mid-1904 the engine was rebuilt with new cylinders each served by slide valves. The original Stephenson valve gear was retained for the inside cylinders, but the outside valves were worked by Walschaerts gear in its first manifestation on a British main line locomotive. No. 271 was rebuilt yet again, in July 1911, but with inside cylinders only and superheated at the same time, and it lasted in this form until withdrawal in 1936.

A second batch of ten 'standard' 4-4-2s built in mid-1903 was preceded six months earlier by No. 251 with its revolutionary large boiler whose diameter was boldly increased from 4ft 8in to 5ft 6in, matched to a *wide* firebox affording a very generous grate area of no less than 31 sq ft. This boiler would indeed be the making of the ninety engines that later followed, but contrary to popular belief, the large firebox was *not* American-inspired, but had been suggested by Ivatt's then works manager, Douglas Earle Marsh. He had previously occupied a similar post on the GWR at Swindon, being there in 1888 while the latter was building the final three 7ft gauge 4-2-2s essentially to Daniel Gooch's classic 1847 Iron Duke design. It was the unusually large fireboxes of those engines, judged by contemporary standards, that were the *true* inspiration for those of the

Great Northern Railway large-boilered 4-4-2 No. 288 (1905). *Former Ian Allan Archive*

London Brighton & South Coast Railway superheated Class H2 4-4-2 No. 421 (1911). *Former Ian Allan Archive*

legendary Ivatt large-boilered 4-4-2s. The final batch of the latter were built in 1910 and provided with eighteen-element superheaters and piston valves from new. Larger twenty-four-element installations soon became the standard from 1912. However, the class as a whole did not achieve its full potential until after considerably larger thirty-two-element superheaters giving very high steam temperatures were installed after 1923 by the LNER, when much larger 4-6-2s were taking over.

The frames of the 1903 batch of GNR small-boilered 4-4-2s had actually been redesigned so as to be readily convertible to take the new large boiler, although this did not occur. Between 1903 and 1906 five more English and Scottish railways would also adopt the 4-4-2, yet only one of these would take up the possible option of a wide firebox. Significantly, the exception was the LBSCR in 1905 under Douglas Earle Marsh, who was now its recently appointed locomotive superintendent.

Unsurprisingly, the Brighton 4-4-2s very closely resembled their GNR antecedents, and they were also even built in Yorkshire, in Leeds by Kitson & Co. to amended Doncaster drawings. One significant difference was the provision of larger cylinders in which the piston stroke was increased from the rather short 24in of the GN engines to the currently more customary 26in. Six additional engines fitted with superheaters, and having improved aesthetics as regards their running boards, were later built by the LBSCR itself at Brighton Works in 1911. The original 1905 4-4-2 series was not superheated until as late as the mid-1920s by the Southern Railway.

In October 1901 an inside-cylinder 4-4-2 with 37 tons adhesive weight was outlined by the Caledonian Railway in Glasgow at St Rollox, but later abandoned in favour of a 4-6-0 with 55 tons resting on its coupled axles, which was unveiled in March 1903 as the 49 class. Also in early 1903, in northern England a 4-4-2 was under design by the North Eastern Railway at Gateshead, while 'alternative' 4-4-2s and 4-6-0s were simultaneously under way *for* the Great Central in Manchester.

The NER 4-4-2 appeared first, in early November 1903, prompted by a recent visit by NER 'top brass' to the USA, where the 'Atlantic' was currently very much in vogue. The initial diagram was prepared and approved in January, barely a month after the appearance of GNR No. 251, and the same 5ft 6in boiler diameter was probably rather more than a mere coincidence. This compared to only 4ft 9in on the S and S1 4-6-0s, whose grate area was increased from 23 to 27 sq ft, but the 4-4-2s' adhesive weight was inevitably inferior at only 40 tons, which was nevertheless the highest yet on a British *four*-coupled locomotive. The 4-4-2 design had been requested by the CME, Wilson Worsdell, on his return from the USA, and was prepared by and under the direction of George

Heppell while the chief draughtsman, Walter Smith, was absent on extended sick leave.[1] The 5ft 6in diameter boiler, subsequently became virtually standard on NER tender locomotives designed after 1902. It also effectively became Heppell's personal trade mark, in later years often also in association with three cylinders, prior to his retirement in 1919.

Only the following month the first 4-4-2 and 4-6-0 express passenger engines were delivered by Beyer, Peacock & Co. to the Great Central in December 1903, having the same boiler and mechanically speaking being very similar other than having respective adhesive weights of 37 and 55 tons. Only a few weeks later, in March 1904, the GCR unhesitatingly placed an urgent order for five more 4-4-2s, to be provided with the slightly deeper firebox such as had originally been proposed by the builder. It is highly significant that it was one of these engines, No. 267, rather than one of the two earlier 6ft 9in 4-6-0s with their cramped ashpans, which when only three months old would make a truly remarkable *through run* of 374 miles from Manchester to Plymouth in October 1904. It departed Manchester London Road at 11.30pm on 28 October, reaching Plymouth at 9.50am the following day. With a GWR pilotman on board, the GC driver was not familiar with the road beyond Oxford. No. 267 departed Plymouth at 12.03am on 30 October, this time fuelled with Welsh coal, and reached Manchester almost ten hours later, at 9.57am – a truly remarkable performance by the engine crew.

Imported French-built four-cylinder compound 4-4-2, GWR No. 102 *La France* (1903), seen making a spirited exit from Paddington on a down express, when newly in service and painted black. *Former Ian Allan Archive*

[1] Heppell, G., *North Eastern Locomotives: A Draughtsman's Life*, North Eastern Railway Association, 2012, p. 13.

Risen to the challenge, GWR four-cylinder 4-4-2 No. 40 (1906), seen when new prior to being named *North Star*. Later rebuilt as a 4-6-0 in 1909, and into the Castle class twenty years later when it also lost its unique valve gear, this locomotive was not finally retired until 1957, after a working life of just over fifty years. *Former Ian Allan Archive*

At this time it was widely felt that four- rather than six-coupled engines were more free running, just as only a few years earlier 'singles' were considered to be less inhibited than four-coupled locomotives. Also, the train loadings on the GCR's new London Extension, upon which they would be employed, were only modest. The 4-4-2 option also possessed a somewhat greater aesthetic appeal (the two 4-6-0s had also been delivered painted in lined black rather than green). The 'Atlantics' were quickly dubbed Jersey Lillies after the renowned contemporary actress and the King's former mistress, Lily Langtry. Ironically, it had originally been proposed to name the two 4-4-2s *King Edward* and *Queen Alexandra*! Arguably, the Great Central 4-4-2s epitomised the Edwardian steam locomotive more than any other locomotive type.

Ordinarily the Great Western Railway would probably not have built any 4-4-2s of its own accord, but it did so as a consequence of its purchase of three French-built de Glehn compound 4-4-2s for comparative purposes. Unlike the later British-built compound 4-4-2s, the interest here would have been the high power to weight characteristics of the French locomotives, rather than seeking fuel economy (although this had been the driver of the very extensive development of the compound locomotive in France, as described in Chapter 9). In other words, these offered the capability of designing a more powerful locomotive within given weight constraints. The first of these, GWR No. 102 *La France*, entered service in October 1903, soon followed by the second taper boiler Saint No. 171 with its boiler pressure increased from 200 to 225lb to make for more valid comparisons.

For the same reason, after being named *Albion* in February 1904 No. 171 was also later rebuilt as a 4-4-2 in October 1904. Thirteen more 4-4-2s were then built new to this pattern during 1905, when two further, larger French compound 4-4-2s arrived. These in turn prompted Churchward to build a four-cylinder simple 4-4-2, No. 40,

which appeared in April 1906 and was quickly named *North Star*. Scant evidence, if any, of actual road testing by the GWR of the French 4-4-2s appears to survive, but Churchward was evidently happily convinced that his engines were superior to the French compounds, moreover with the virtue of having less complication, by achieving his objective of developing a drawbar pull of 2 tons at 70mph (equal to 836 drawbar horsepower). At the same time, he resolved to build 4-6-0s rather than 4-4-2s in future on account of their greatly superior adhesive weight, i.e. around 55 rather than only 39 tons, particularly in view of the gradients out west in Cornwall and at Plymouth. Early on No. 40 *North Star* had proved too powerful for its limited adhesive weight by slipping badly and bending its original I-section coupling rods. It was converted to become a 4-6-0 in November 1909, immediately after a total of thirty production four-cylinder Star class 4-6-0s derived from it had been put into service since February 1907.

As to the three French compounds, their original boilers were replaced by Swindon Standard No. 1 taper boilers between 1907 and 1916, and they would later be withdrawn between 1926 and 1928. A lasting legacy, however, was the bequest of their smooth-running French bogie design that became incorporated into Swindon standard practice on hundreds of 4-6-0s. Also, the distinctive 'swept back' positioning of their outside cylinders was emulated in the three successive classes of four-cylinder 4-6-0s totalling 260 engines that were built at Swindon between 1907 and 1950.

In April 1905, John F. McIntosh on the Caledonian Railway had also briefly toyed with the idea of a four-cylinder compound 4-4-2 on de Glehn lines, but quickly abandoned it. Later that same year he obtained actual sanction to build five inside-cylinder 4-4-2s. Although these were not built either, they would have otherwise closely resembled the magnificent 903 class 4-6-0s that first appeared in May 1906.

If the 4-4-2 appeared to be fated on the Caledonian Railway, it did finally appear on another Scottish railway in July 1906, on the slightly larger but rather less glamorous North British Railway, which by 1905 was badly lagging behind the former in terms of locomotive prestige. The turbulent debut of these particularly striking locomotives was chronicled with reference to contemporary primary sources by the Scottish railway historian, John Thomas.[2] Why these were built as 4-4-2s rather than 4-6-0s, in view of some of the gradients that they would encounter, has been the subject of much debate over many years. The sinuous curves on the Edinburgh–Carlisle main line (the Waverley Route) have been suggested as being the explanation, but on the other hand this route was equally characterised by sustained severe gradients, e.g. 7 miles at 1 in 75! By adopting an Atlantic the NBR did show solidarity with its English East Coast partners, the GNR and NER, while demonstrating a pleasing independence from the Drummond philosophy that still dominated on the rival Caledonian. This was further emphasised by the

very unusual employment by the NBR of *outside* cylinders, and the omission of the hitherto traditional Stroudley/Drummond smokebox wingplates.

The fourteen 4-4-2s, built by NBL, were primarily intended to work the new Edinburgh–Aberdeen block trains, and it was initially recorded in early November 1905 that these would be three-cylinder compounds, probably similar to but larger than the very recent Robinson compound 4-4-2 on the Great Central.[3] However, this statement had almost immediately been revised instead to become two-cylinder simples. The NBR 4-4-2s as built could be said to have been virtually standard Robinson GCR 'Atlantics', whose engine wheelbase they replicated *exactly*, scaled up to NER Class V proportions with 20in by 28in cylinders, and likewise served by a 5ft 6in diameter (but now Belpaire) boiler pressed to 200lb, via 10in diameter outside admission Smith-pattern piston valves. The NBR was, of course, well familiar with the Vs in Edinburgh, while, possibly significantly, during 1905 a batch of 4-4-2s for the Great Central was built in Glasgow by NBL. John Thomas suggested that the NBR chief draughtsman, Walter Chalmers, was a close friend of John Robinson.

[2] Thomas, J., *The North British Atlantics*, David & Charles, 1972.
[3] *Ibid*, p. 37.

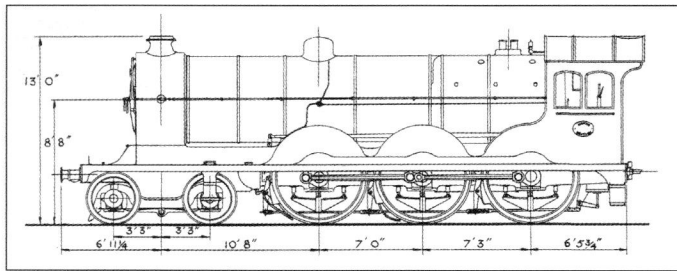

13' 0"
8' 8"
3' 3" 3' 3"
6' 11¼" 10' 8" 7' 0" 7' 3" 6' 5¾"

*Left:* Outline drawing by W. D. Stewart, based on the Cowlairs general arrangement drawing, of the proposed NBR inside-cylinder express passenger 4-6-0 that was designed in 1907 in the almost immediate wake of the 4-4-2s' arrival.

*Below:* North British Railway 4-4-2 No. 870 *Bon Accord* awaits departure at Edinburgh Waverley Station with a train for Perth, c. 1909. *Rail Archive Stephenson*

Amid a number of almost immediate alleged operational problems, the new 4-4-2s were very soon accused by the NB civil engineer, James Bell, of damaging the track. Vincent Raven from the NER and Henry Ivatt on the GNR were each called in to make independent reports.[4] Twenty years later the NB 4-4-2s' hammer blow at 5rps was determined by the Bridge Stress Committee to be 8.0 tons at 72mph, although this was rather less than the corresponding figure of 9.9 tons for the ex-Great Central 4-4-2! This particular criticism had soon died away, however, and no detectable reductions in the size of the balance weights on the driving wheelsets of the two subsequent batches of NBR 4-4-2s, built in 1911 and 1921, were evident.

The furore when the 'Atlantics' first appeared prompted Cowlairs in mid-1907 to design an inside-cylinder 4-6-0, which would have born a close resemblance to the recent Caledonian Railway 903 class, but for its Belpaire firebox and side window cab.[5]

Within the NBR locomotive department there had also been questions as to whether the Atlantics should have been built as compounds as initially proposed. To try and resolve this a Midland Railway Deeley compound 4-4-0, No. 1032, and both of the North Eastern compound 4-4-2s, Nos 730 and 731, ran over the NBR's Edinburgh–Carlisle route during 1908. However, the results did not impress the NBR sufficiently to encourage it to go compound after all. Then, two years later in October 1910, by arrangement with the LNWR, a North British 4-4-2 made a single run between Preston and Carlisle, pitted against a Whale Experiment 4-6-0, in the process consuming 71lb of coal per mile, compared to only 58lb by the somewhat smaller LNW 4-6-0 running in its own familiar territory.

After all this circumspection six more 4-4-2s were ordered from Robert Stephenson & Co. in February 1911. Although slightly modified, unlike the second batch of Brighton 4-4-2s that would have been authorised at almost the same time, these were not superheated. Although superheaters were later fitted to the 1906-built NBR 4-4-2s between 1915 and 1921, the 1911 series were not similarly dealt with until as late as 1923–25 under LNER auspices. This made for an interesting contrast with the NER non-superheated Z 4-4-2s that had been built almost simultaneously, and which were very quickly superheated ten years earlier than the NB engines.

All British 4-4-2s were ultimately superheated, bar GNR No. 292 and all those on the Lancashire & Yorkshire Railway. Rather surprisingly though, some of the latter could nevertheless be said to have been early pioneers of superheating, in that the last of the 1899 series, No. 797, and the last five of the 1902 series, Nos 1420–4, were each fitted when built with a steam dryer in the smokebox. To accommodate this the front tubeplate was recessed 3ft into the boiler barrel, thereby reducing the fire tubes to a length of only 12ft and the total evaporative heating surface by 19 per cent. This equipment elevated the steam temperature by 95°F and was said to have reduced coal consumption by 3½lb per mile, while John Aspinall claimed that the engines so fitted also ran more freely. Such was not removed from No. 1424 until 1917, so that overall the device had the unusually long currency of nearly twenty years. It is surprising, in view of Horwich's early interest in superheating, that true superheaters were never fitted to any of the 4-4-2s, despite the fact that many of them later received new boilers after 1906.

In January 1907 the Great Central turned out its final Atlantic. In fact, the GCR did not turn out any more express passenger locomotives until December 1912, and later in August 1913 it introduced the first of ten large superheated Director class 4-4-0s that were more than equal to the earlier 4-4-2s in terms of haulage capacity and adhesive weight.

The Great Northern Railway built no 4-4-2s after 1910, by which time it had a substantial fleet of 116 Atlantics, ninety-three of these with wide fireboxes. It is known, however, that in early 1914 Nigel Gresley was contemplating a new 4-4-0 with a 20-ton axle load, which therefore should have been comparable with the GCR Director.

The ultimate British 4-4-2 design was the elegant North Eastern Railway three-cylinder Class Z, introduced in 1911, of which a total of fifty was built up to 1918, although the final twenty had actually been authorised in April 1914. Like the Midland Compound 4-4-0s, to which they were distantly related via the unique rebuilt NER 4-4-0 No. 1619, these boasted three pairs of eccentrics on the single throw crank axle, which would not have endeared them to shed maintenance staff. A skilled designer that he was in many other respects, it has to be said that the chief draughtsman, George Heppell, like his Swindon counterparts probably would rarely have ventured out of the relative comfort of the drawing office. In Heppell's defence it also has to be said that while at Robert Stephenson & Co. at the age of 16 he had requested to undergo a full apprenticeship in the works before returning to the drawing office, but this had been refused.[6]

How, then, did the 'greyhound' Z class actually compare in traffic when compared with the 'bulldog' V from which it had been directly derived? By early LNER days, owing to their lighter and mutually better balanced moving parts, the three-cylinder Zs were officially scheduled to undergo heavy repairs at 75,000-mile intervals, compared to only 65,000 miles for the two-cylinder V. In 1924 at least, the actual differential was reported to have been somewhat greater at 73,000 and 58,000 miles respectively.[7]

[4] Ibid, pp. 79–92.

[5] Barnes, R., Locomotives That Never Were, Jane's, 1985, an artist's impression of the proposed NBR 4-6-0, is featured on p. 16.

[6] Heppell, G., p. 4.

[7] Locomotives of the LNER, Part 3A, Tender Engines Classes C1 to C11, The Railway Correspondence & Travel Society, 1979, p. 95.

Superheated North Eastern Railway three-cylinder Class Z 4-4-2 No. 729 stands on the King Edward Bridge, spanning the River Tyne between Newcastle and Gateshead. Probably taken when the engine was brand new, with original brass-capped chimney, in September 1911. *R. J. Purves/Rail Archive Stephenson*

Whichever, both sets of mileages were considerably below the corresponding figures for the four-cylinder Star and two-cylinder Saint 4-6-0s on the GWR (see Chapter 7). In addition, while on the Z there was one additional cylinder and its associated set of valve gear to attend to when repairs fell due, if just one cylinder among the three required replacement, then the complete monobloc casting, which embraced all three cylinders, needed to be renewed. On the later evidence of the Bridge Stress Committee, compared to the V the Z was a little kinder to the track, etc, on which it would have inflicted only 1.9 tons hammer blow at 5rps (73mph), as against 3.3 tons by the V.[8] The highest recorded line speed by a V was 75mph, and 82mph by a Z.

It has only rarely been possible to compare the coal consumption of otherwise similar two- and three-cylinder locomotives, but here the Zs also scored, no doubt on account of their softer and more even exhaust characteristics. NER figures for January–December 1920 plus July–December 1921 (to eliminate the effects of the miners' strike in early 1921), by which time all that company's 4-4-2s were superheated, gave 65.6lb/mile for the V, yet only 50.6lb/mile for the Z, a notable 23 per cent reduction.

Although almost fifteen years elapsed between the emergence of the first two-cylinder and final three-cylinder 4-4-2s, both groups were retired between 1943 and 1948, the demise of the more recent Zs being hastened by the rapid influx of new Thompson LNER B1 4-6-0s immediately after the end of the Second World War. Nevertheless, analysis of the total life mileages available for thirteen V/V1s and thirty-one Zs shows that the latter achieved a higher figure by about 10 per cent, and proportionately an even higher average *annual* mileage, although latterly both classes had predictably gone into a sharp decline after 1938:

### Table 12 Comparison of average lives and mileages of NER Class V and Z 4-4-2s

|  | 2-cylinder V/V1 | 3-cylinder Z |
|---|---|---|
| Built | 1903–10 | 1911–17 |
| Average life mileage | 1,027,505 | 1,129,781 |
| Average working life | 38y 10m | 32y 10m |
| Average annual mileage | 26,579 | 34,392 |

[8] The Bridge Stress Committee Report attributed this surprisingly low value to nineteen of the twenty V/V1s, but what was surely a more characteristic value of 5.9 tons to the remaining engine. From the Gateshead drawing register it is apparent that very shortly before his promotion to chief draughtsman George Heppell had designed new balance weights for the Class V engines allowing for 50 and 66 per cent reciprocating balance. This was possibly prompted by the very recent arrival of the potentially smoother running four-cylinder compound 4-4-2s. It is not known if either of these was implemented, or what the original balanced percentage had been.

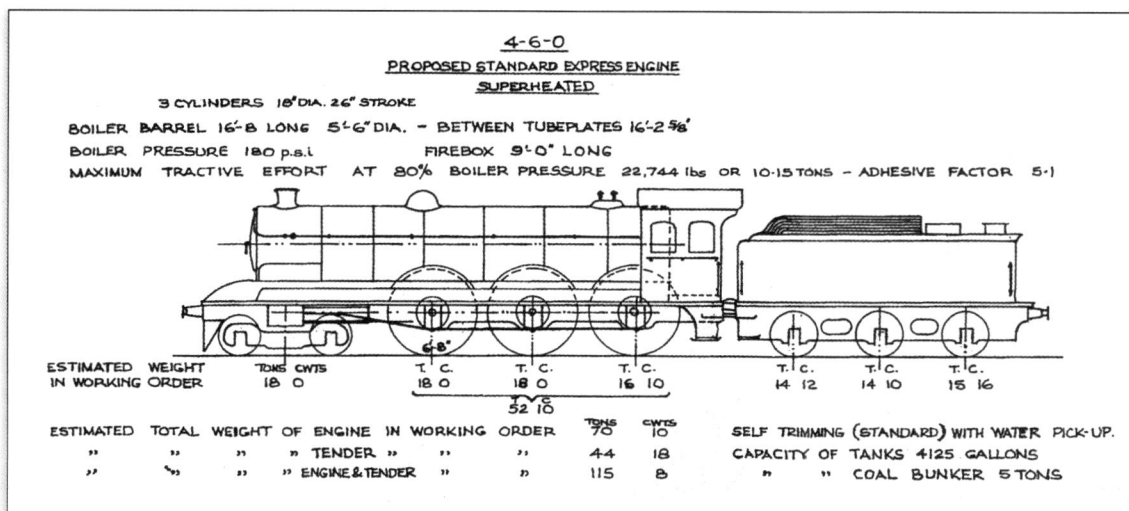

Diagram of proposed NER 3-cylinder 4-6-0 express passenger 4-6-0, 1918.

## Proposed national standard locomotives

During 1917–18 the ARLE forwarded to its members a request from the government to put forward outline proposals for possible future national standard locomotives, for construction after the current conflict had ended. In response, the NER submitted diagrams for highly interchangeable three-cylinder heavy mineral 0-8-0s and mixed traffic 4-6-0s, which it then promptly went on to build for itself as classes T3 and S3 respectively, together with an express passenger version of the latter with 6ft 8in diameter coupled wheels, which would have become NER Class S4.[9] This would virtually have amounted to a six-coupled version of the Z 4-4-2, but with larger 18in in place of 16½in diameter cylinders. Its estimated adhesive weight was 51½ tons compared with the 40¾ tons of the 4-4-2. However, in reality had this 4-6-0 been built this would probably have fallen not far short of 60 tons, as the adhesive weight actually came out at 7¼ tons over the estimate at 58¾ tons for the very similar S3 mixed traffic 4-6-0 when that was completed. The latter's maximum axle load was 20 tons, a figure that fortunately had already long been acceptable on the NER, although it had been estimated to be only 17½ tons. A much smaller discrepancy elsewhere that had also involved a 4-6-0 a few years earlier in Scotland had resulted in the locomotive superintendent's resignation (see Chapter 10).

These significant underestimates are surprising given both 4-6-0s' very close affinity with the well-established Class Z 4-4-2 (79 tons), while on the other hand the estimate for the more compact three-cylinder 0-8-0 at 71 tons had proved to be extremely accurate. The first three-cylinder mixed traffic 4-6-0s and five large 0-8-0s were completed at Darlington Works as soon as late 1919, although the actual justification for building the T3 has

remained a mystery, which was further compounded by the LNER building another ten of these engines in 1924.

One can only speculate as to why the projected NER S4 4-6-0 was not built, although immediately after 1918 building new goods and mixed traffic locomotives widely tended to be the priority. As early as 1912 new Class Z 4-4-2s were already hauling 545-ton trains between York and Newcastle, and by early 1920, on account of their reduced frequency, the heaviest express passenger loadings between London and Edinburgh were regularly approaching 600 tons. In response the Great Northern Railway under Nigel Gresley began to design a powerful three-cylinder 4-6-2. On 30 March 1922, the same day that GNR No. 1470 *Great Northern* was officially completed at Doncaster, the North Eastern directors also sanctioned the construction of two rather different three-cylinder 4-6-2s, which in effect amounted to a 'stretched Z' with a semi-wide firebox. The first, NER No. 2400, was completed only a little over seven months later at Darlington Works. At 101½ tons this weighed 4½ tons over the original estimate, and 9 tons more than the GNR engine, with which it nevertheless shared, officially at least, the same adhesive weight of 60 tons. On 1 January 1923, of course, both companies, together with the North British and others, would become absorbed into the new London & North Eastern Railway.

## 4-4-2 finale

In the contemporary context it is rather surprising, therefore, that the North British Railway took delivery of two additional 4-4-2s from NBL as late as 1921. Although these would indeed prove to be the last British 'Atlantics'

9 Heppell, G., p. 19.

actually to be built, at the time of its legal demise it is known that the NBR had forward plans to build another five of these.

To summarise: when compared with a 4-4-0 or 4-6-0, on the 4-4-2 the low proportion of its adhesive weight in relation to engine weight (tender excluded) soon began to became a disadvantage as express passenger train weights and speeds continued to increase. It was a practical reason why relatively few 'Atlantics' were actually built over a fairly limited period and by only half of the fourteen major British railway companies. An excellent example of this deficiency was to be found on the North Eastern Railway, on whose R1 4-4-0 the adhesive weight amounted to 70 per cent, while on the S3 4-6-0 (to which the proposed S4 passenger engine would have very closely approximated) it was slightly more at around 75 per cent of the engine weight. On the other hand, on the almost legendary Z 4-4-2 *when static* it was only 50 per cent at best, but could potentially diminish when the engine was pulling hard on a stiff gradient through weight transference onto the unproductive trailing carrying axle. On the short-lived GWR 4-4-2s (as such), both with two and four cylinders, the adhesive weight had amounted to a slightly higher proportion at 55 per cent. That said, despite its inherent limitations and comparative rarity, the British 4-4-2 tender engine nevertheless operated for precisely sixty years, between 1898 and 1958.

In Britain, as elsewhere, the Atlantic was essentially an Edwardian phenomenon.[10] The last example to remain in active service, was former LBSCR No. 424, new in August 1911, which was withdrawn from service in April 1958 after a commendably long working life of almost forty-seven years.

[10]An excellent worldwide survey of the 4-4-2 tender engine is provided in: Hennessey, R. A. S., *ATLANTIC, The Well Beloved Engine*, Tempus Books, 2002.

# 7
# 4-6-0 Mania

T he 4-6-0 proved to be uniquely versatile under British operating conditions, and it rapidly gained popularity. This success could be attributed to three favourable factors:

1. The tolerance of fairly high axle loadings by contemporary European standards.
2. The high proportion of its weight that was available for adhesive purposes.
3. The availability of good-quality coal that could be burned in a narrow firebox, accommodated between traditional plate frames.

In December 1900 there were only thirty-four 4-6-0s as yet at work on Britain's railways, divided between the Highland (with twenty-one), the North Eastern (eleven), and the Great Western (two). The two GWR engines, Nos 36 and 2601, were very short-lived, double-framed oddities attributed to William Dean. After 1900 the 4-6-0 proliferated rapidly in numerical strength, while widely differing as regards its boiler, cylinder and valve gear configurations. By December 1914 a total of almost eight hundred 4-6-0s were in service in Britain on nine major British railways, and also on the Highland Railway, which had pioneered the type in 1894.

The prototype for a class that ultimately totalled 130 locomotives, London & North Western Railway four-cylinder 4-6-0 No. 2222 *Sir Gilbert Claughton* (with indicator shelter attached), is inspected by the LNWR Locomotive Committee at Crewe Works on 7 March 1913. Sir Gilbert himself, the company chairman, is seen at the regulator, while the CME, C. J. Bowen Cooke, stands directly below the number plate. *Science and Society Picture Library*

## Table 13 British 4-6-0 classes introduced between 1901 and 1914

| Year | Passenger | | | Mixed traffic | Fast goods |
|---|---|---|---|---|---|
| | Inside cylinders | Outside cylinders | Four cylinders | | |
| 1901 | | | | | |
| 1902 | CR 55<br>5ft 0in/Step. vg | GWR No. 100<br>6ft 8½in/Steph. vg | | GCR 8 (O)<br>6ft 0in/Steph. vg | |
| 1903 | CR 49<br>6ft 6in / Step. vg | GWR No. 98<br>6ft 8½in/Steph. vg<br>GSWR 381<br>6ft 6in/Steph. vg<br>GCR 8C<br>6ft 9in /Steph. vg | | | LNWR 1400<br>5ft 2½im/Joy vg<br>(four-cylinder<br>compound) |
| 1904 | | | | | |
| 1905 | LNWR Expt<br>6ft 3in /Joy. vg | | LSWR F13<br>6ft 0in/Wals. & Steph. vg | | |
| 1906 | CR 903<br>6ft 6in/Steph. vg<br>CR 908<br>5ft 9in/Steph. vg | | | GCR 8F (O)<br>6ft 6in/Steph. vg | CR 918 (I)<br>5ft 0in /Steph. vg<br>GCR 8G (O)<br>5ft 3in/Steph. vg<br>LNW 19in (I)<br>5ft 2½in/Joy vg |
| 1907 | | | GWR Star<br>6ft 8½in/Wals. vg<br>LSWR E14<br>6ft 0in/Wals. & Steph. vg | | |
| 1908 | | | LSWR G14<br>6ft 0in/Wals & Steph. vg<br>LYR 1506<br>6ft 3in /Joy vg | | |
| 1909 | | | | | |
| 1910 | | | LSWR P14<br>6ft 0in/Wals. & Steph. vg | | |
| 1911 | LNWR PoW*<br>6ft 3in/Joy vg<br>GER 1500*<br>6ft 6in/Steph. vg | GSWR 128*<br>6ft 6in/Steph. vg | LSWR T14<br>6ft 7in/Wals. vg | NER S2 (O)<br>6ft 1in/Steph. vg | |
| 1912 | GCR 1*<br>6ft 9in/Steph. vg | | | | |
| 1913 | | | LNWR<br>'Claughton'*<br>6ft 9in/Wals. vg | GCR 1A* (I)<br>5ft 7in/Steph. vg<br>CR 179* (I)<br>5ft 9in/Steph. vg | |
| 1914 | | | | LSWR H15 *(O)<br>6ft 0in/Wals. vg | |

Key: (I) inside cylinders, (O) outside cylinders, e.g. 6ft 3in coupled wheel diameter.
* = Superheated; Steph. = Stephenson; Wals. = Walschaerts; Joy = Joy valve gear.

After twenty years since first introduction, the great majority of 4-6-0s had large-diameter coupled wheels (6ft 3in to 6ft 9in) for express passenger service, usually without any notable success with the exception of on the GWR. A much smaller number of 4-6-0s, however, had intermediate-sized coupled wheels (5ft 6in to 6ft) and were what would increasingly be termed mixed traffic engines, although 'maids of all work' was the popular phraseology at that time, which unconscientiously reflected contemporary social norms. In addition, the LNWR built a total of 200 4-6-0s with 5ft 2½in diameter coupled wheels that were specifically intended as fast goods engines, as were ten similarly on the Great Central with 5ft 3in. There were some anomalies, however. The Caledonian Railway's first 4-6-0s, the 55 class with 5ft diameter coupled wheels (1902), were specifically built for passenger duties on the demanding Callander & Oban line, while a later larger-boilered version, the 918 class (1906), was for fast goods. The Great Central 8F with 6ft 6in diameter coupled wheels (also 1906) was virtually a copy of the earlier express passenger 6ft 9in Class 8C, but was intended to work passenger-rated express *fish* traffic from Grimsby.

The most enduring group of British 4-6-0s ever built could directly trace their origins back to Swindon in January 1901, the first month of the 20th century, when the 63-year-old Victorian era also ended and the new Edwardian period began. While William Dean was as yet

A fine study of the pioneer Churchward 4-6-0 No. 100 newly at work, as completed in February 1902, and before being named firstly simply *Dean* in June 1902, and subsequently *William Dean* only five months later. The unique parallel boiler was replaced after only sixteen months in June 1903.
*The Great Western Trust*

still officially in charge at Swindon, a very simple composite diagram was prepared showing six proposed radically new standard locomotive types that was laid out as below:[1]

| 2-8-0<br>(4ft 7½in) | 2-6-2T<br>(5ft 8in) |
| --- | --- |
| 4-6-0<br>(5ft 8in) | 4-4-2T<br>(6ft 8½in) |
| 4-6-0<br>(6ft 8½in) | 4-4-0<br>(6ft 8½in) |
| (Coupled wheel diameter) | |

Each type featured mutually standardised 18in diameter outside cylinders having the unusually long piston stroke of 30in, and showed parallel domeless boilers having a raised Belpaire firebox casing. Such had made its first appearance in October 1899 on the 5ft 8in Bulldog 4-4-0 No. 3352 *Camel*, and soon afterwards in 1900 on the prototype Aberdare double-framed 2-6-0, and Dean 2-4-2 tank engines, later to be followed on the City 4-4-0s in early 1903. Soon after this it would be superseded by the taper boiler. In the event, only a single example of just one of the above six types actually appeared in the basic form indicated with the early pattern of Belpaire boiler, i.e. the 6ft 8½in 4-6-0 as GWR No. 100 in February 1902. (Its 5ft 8in counterpart, then primarily envisaged with the gradients in the West Country in mind, on the other hand would not finally emerge in spirit and in superheated form until thirty-five years later as the Charles Collett Grange class in 1936.) Undoubtedly designed under the guidance of George Churchward, who would very soon take over from William Dean, the *ensuing* concepts would remain enshrined in regular production at Swindon Works until 1950, although incorporating two highly distinctive features that had not yet been envisaged in January 1901:

■ The unique parallel boiler first fitted to No. 100 was 5ft in diameter and contained 287 2in diameter fire tubes, which was mated to a 5ft 6in wide outer firebox casing. This boiler was replaced after only sixteen months in June 1903. It would appear that the new domeless Belpaire boilers in general suffered from broken firebox stays and cracks at the junction of the outer firebox casing and the boiler barrel on account of poor water circulation. Before the original boiler on No. 100 had scarcely entered service a more sophisticated successor had already been designed. By comparison, the taper boilers were expanded to 5ft 6in maximum diameter at the firebox end, yet contained only 250 2in tubes. This resulted in a comparative reduction in 'tube density' of 28 per cent, thereby making for greatly improved water circulation and therefore steam-raising capacity. The firebox itself was also more sophisticated in that it incorporated curved sides both inside and outside, and was further slightly tapered in its own right, narrowing towards the cab. Although in early Churchward boilers the taper was confined to the rear barrel ring (termed half cone), in 1906 the taper had been altered so as to extend uniformly throughout both rings (full cone), not only in the relatively long Standard No. 1 but also in the shorter Nos 2 and 4 boilers.

■ The cylinders on No. 100 (which surprisingly apparently lasted throughout the thirty-year life of the locomotive, until finally worn out) were also unique, and contained short-travel piston valves that were originally of only 6½in diameter, although this was later increased to 7½in. Only then really coming into vogue, piston valves permitted the provision of considerably

---

[1] This seminal document was discovered by Michael Rutherford, who is an authority on George Churchward and his work. It is reproduced on p. 9 of his book *'Halls', 'Granges' & 'Manors' at Work*, Ian Allan, 1985. This complements his earlier work, *'Castles' & 'Kings' at Work*, Ian Allan, 1982, which is also recommended reading.

enlarged steam ports, at a time when locomotive cylinders were increasing in size. When compared with No. 100, the steam port area in the completely redesigned cylinders, now provided with 10in diameter piston valves, was increased by 65 per cent, and those of the exhaust ports by no less than 170 per cent. Designed in the contemporary American style, these were to be cast integral with half of the smokebox saddle, the two identical castings being bolted together down the vertical centre line. Being set horizontally, these castings were therefore not handed, although while the cylinder dimensions remained uniform, the radius of the smokebox saddle would necessarily have to change depending on the variant of taper boiler fitted, i.e. whether Standard No, 1, No, 2 or No. 4. The maximum travel of the piston valves was increased to almost 6in, against a contemporary norm of only around 4in, while their lap was 1⅝in as against a typical 1in or sometimes even less.

In truth, No. 100 can have had little *direct* influence on the locomotives that followed, because the Swindon drawing registers record that both the respective arrangement drawings for the new taper boiler *and* the improved cylinders were dated March 1902. These first appeared together on 4-6-0 No. 98 one year later in March 1903, and three months after that on the prototype 2-8-0 No. 97

in June. Also fitted with smaller taper boilers, but with similar cylinders were 2-6-2T No. 99 in September 1903, the first outside-cylinder County 4-4-0s in May 1904, and their 4-4-2T counterparts in September 1905.

Away from the Pennsylvania Railroad and Great Northern Railway, Belpaire boilers were comparatively rare in North America. However, the new GWR taper boilers had every appearance of having been inspired by the recent products of the Brooks Locomotive Works in the USA, despite that builder having been referred to by its detractors as Cold Water Brooks, on account of the allegedly poor steaming qualities of its boilers. The distinctive form of the new GWR outside cylinders with their inwardly inclined external steamchests, in association with internal Stephenson valve gear, quite definitely reflected what was then standard practice in the USA. Both of these features would very soon be discontinued in North America as a result of the advent of Walschaerts and other external valve gears. At Swindon, on the other hand, neither would be abandoned in new construction as regards the large two-cylinder locomotives that were built right through to the GWR's demise in 1947, and indeed for a further three years beyond.

The 1903 cylinder and long-travel valve gear designs, as applied to the 4-6-0s in particular, were superb. There has long been much speculation as to whether or not GWR 4-4-0 No. 3440 *City of Truro* actually broke the 100mph barrier on 9 May 1904. Rather less well known is the even more remarkable claim that 4-6-0 No. 2903 *Lady of Lyons* may have hit a speed as high as 120mph while

Perspective drawing of the classic arrangement of long-travel Stephenson valve gear with 'launch'-type expansion links, as applied to Churchward GWR two-cylinder locomotives and their successors. These culminated in the Hawksworth County class 4-6-0 (1945), which had 7½in maximum valve travel.

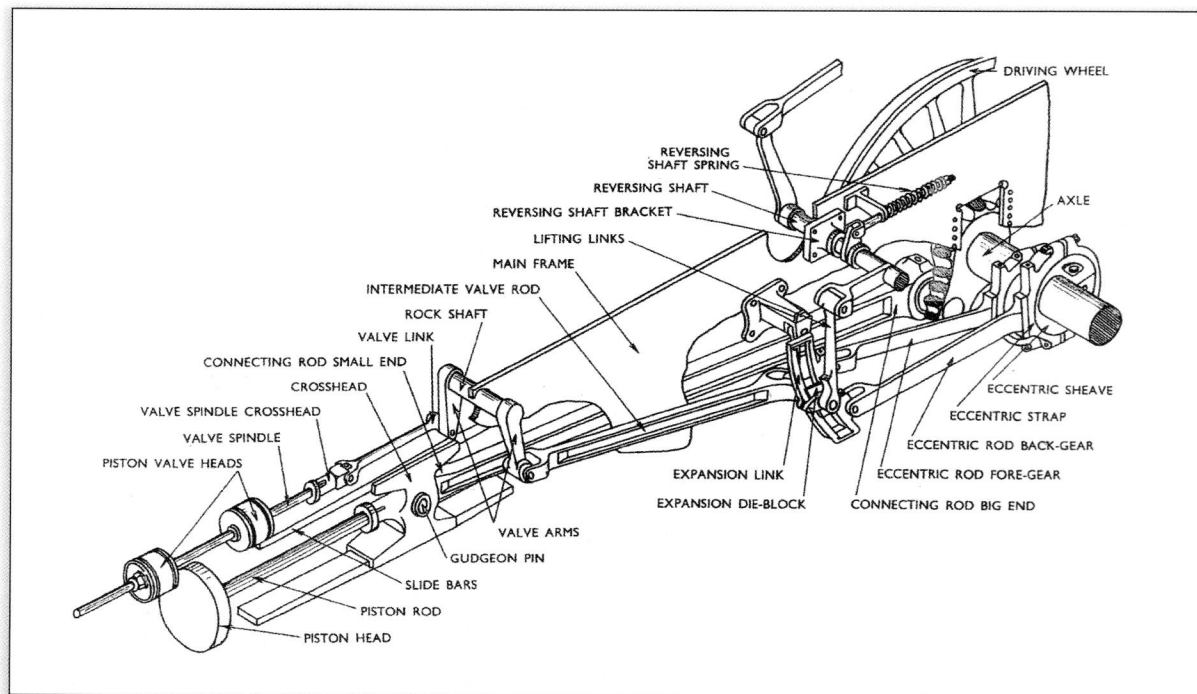

GWR two-cylinder 'Saint' class 4-6-0 No. 2922 *Saint Gabriel* (1907) as built with improved running board styling designed by Harold Holcroft at George Churchward's request following criticism in the press. The narrow chimneys fitted to 'Saints' and 'Stars' at this period were later replaced by a slightly wider variety. *Former Ian Allan Archive*

GWR four-cylinder Star class 4-6-0 No. 4009 *Shooting Star* (1907) as built. *Former Ian Allan Archive*

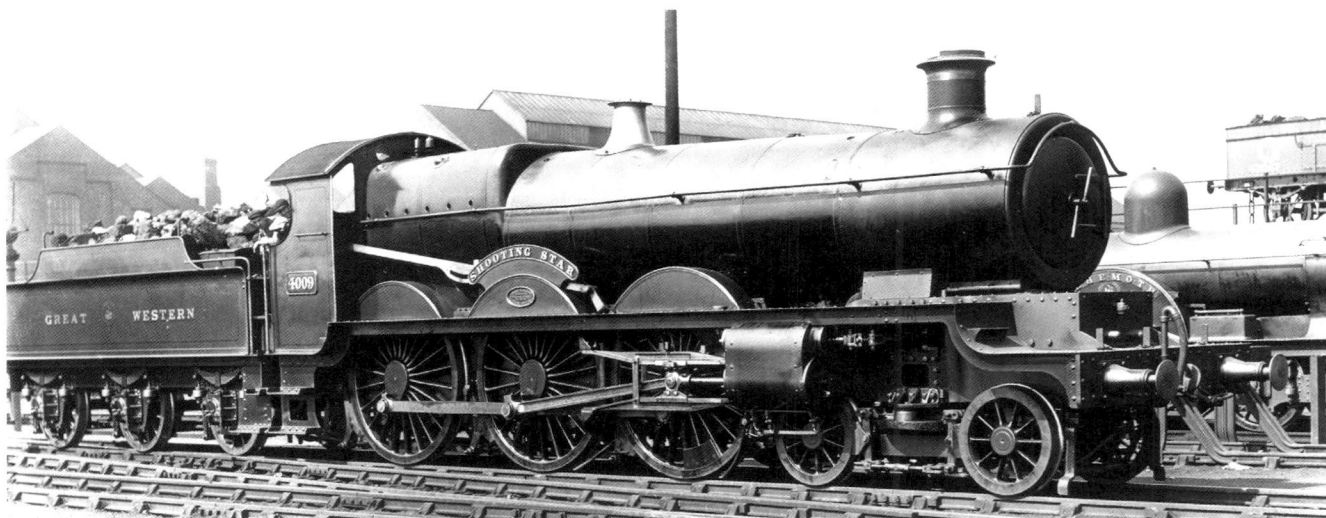

running light engine down a 1 in 300 falling gradient between Badminton and Little Somerford, in the course of trials when brand new in May 1906. No less than Charles Collett, then Assistant Works Manager at Swindon, who sixteen years later would become Churchward's successor as CME, was on the footplate at the time. About twenty-five years later he would admit only that a very high speed had been achieved west of Swindon. As with the earlier 4-4-0 exploit, timing had been via a stopwatch and the observation of mileposts flashing past, and so could inevitably have been subject to significant error.

In a third 4-6-0, No. 171, which was completed the following December, its boiler had been accordingly redesigned to allow the working pressure to be increased from 200 to a then distinctly bold 225lb/sq in, in order to match the 227lb of the newly arrived French compound 4-4-2 No. 102 *La France*. After a few months, No. 171 was also rebuilt as 4-4-2 to make for a still fairer comparison. No. 172, built as a 4-4-2, later followed in February 1905, which was succeeded by Nos 173–8 built as 4-6-0s, and Nos 179–190 as 4-4-2s again, all of which

were turned out between February and September 1905 and given names associated with the novels of Sir Walter Scott. Other than No. 171, which was converted back again to a 4-6-0 in 1907, the 4-4-2s that had been newly built as such were later converted to 4-6-0s during 1912, when they were also renumbered as 2972–90. By this time the GWR two-cylinder 4-6-0s had collectively become known as the Saint class.

These now only awaited the refinement of superheating and top feed, which would be newly incorporated in the course of series production in 1910–11, and the general pattern had been set for nearly five hundred closely related 4-6-0s that would all be built at Swindon Works until the mid-point of the 20th century. From 1907 a four-cylinder version of the Saint, the Star class was introduced, via the solitary four-cylinder 4-4-2 No. 40 built the previous year. However, the latter's unique eccentric-free 'scissors' valve gear was not repeated, and it was replaced by the inside Walschaerts valve gear already designed for *The Great Bear*. While a very similar steam port area to cylinder volume ratio, and the same 1⅝in piston valve lap of the two-cylinder 4-6-0s was maintained, the maximum valve travel of the four-cylinder engines was increased by one full inch. On the minus side, the exhaust steam ducts from the four cylinders to the base of the blastpipe were inevitably longer and less direct on the Stars than on the two-cylinder engines, although these were improved in later batches.

Between 1906 and 1914 Saints and Stars were sometimes built in alternating batches as indicated below, usually at the rate of ten new engines each year. The Stars on average were 18 per cent more expensive than the Saints, on account of their increased complexity and greater weight. It was the weight of the early non-superheated engines that was always quoted, i.e. 75.8 tons. The increased weight of the class when superheated was never officially acknowledged, but it was probably nearly 80 tons.

Saint   GWR Nos 2901–2910 (10) May 1906
Star    4001–4010 (10) February–May 1907
Saint   2911–2930 (20) August–September 1907
Star    4011–4020 (10) March–May 1908
Star    4021–4030 (10) June–October 1909
Star    4031–4040* (10) October 1910–March 1911
Saint   2931–2940* (10) October–December 1911
Saint   2941–2950* (10) May–June 1912
Saint 2951–2955* (5) March–April 1913†
Star 4041–4045* (5) May–June 1913
Star 4046–4060* 15) May–July 1914

* Superheated from new (Swindon superheater), No. 2901 was built with a Schmidt superheater, and was also the first British locomotive to be fitted with a true superheater.

† Nos 2956–60 had also been authorised, but were cancelled in December 1912.

Although both classes used the same Standard No. 1 boiler, the four-cylinder Stars were rated one coach more powerful, which Harold Holcroft attributed to their Walschaerts valve gear, and they were best suited to running heavy, high-speed, non-stop trains, which frequently became lighter en route through the detachment of slip coaches. They were also naturally smoother running, and although locomotive balancing was never a strong point at Swindon, at 5 rps (72mph) a Star would (after a later revision) deliver a hammer blow of only 2.0 tons, compared with 4.8 tons by a Saint. Also, the Stars could run 120,000 miles between heavy repairs, as against only 80,000 miles by the Saints.

Taking the five Stars and five Saints that were built in 1913 as representative samples, of which none of the former were later rebuilt into larger Castles during their long lives, it is interesting to note that, perhaps surprisingly, the respective average working life spans and average official life mileages of the two groups were to all intents and purposes identical:

### Table 14 Life, mileage & cost comparisons of GWR 1913-built Saint and Star class 4-6-0s

|  | Saint (Court sequence) | Star (Prince sequence) |
|---|---|---|
| GWR Nos | 2951–5 | 4041–5 |
| Built | March–April 1913 | May–June 1913 |
| Cost per loco* | £2,650 | £2,974 |
| Withdrawn from service | May 1950–Jul 1952 | Nov 1950–Feb 1953 |
| Average working life | 38 years 8 months | 38 years 7 months |
| Average life mileage | 1,721,317 | 1,772,657 |
| Average annual mileage | 44,524 | 44,652 |

* Engines only, the tenders would have cost approximately £500 each

In 1919 George Churchward proposed to up rate the Saints and the Stars (and the 28XX 2-8-0s) by fitting them with the new Standard No. 7 taper boiler that was currently under design for a new mixed traffic 2-8-0. Compared to the Standard No. 1 boiler, grate area would have been increased from 27.1 to 30.3 sq ft, while the new boiler would weigh about 2¾ tons more. However, in addition to possible weight problems it was discovered that the firebox 'shoulders' of the new boiler could not be accommodated within the GWR loading gauge in association with 6ft 8½in coupled wheels. The Collett Castle class, introduced in 1923, amounted to a slightly scaled down version of Churchward's proposed 'Super Star', and would remain in production until as late as 1950.

Notwithstanding past glories, and undeniably a highly successful design, the construction of moderately sized 4-6-0s having the luxury of *four* cylinders, particularly after 1939, and essentially to a basic design that pre-dated 1914, was a distinct anachronism, as also was the retention of inside Stephenson valve gear in new two-cylinder 4-6-0s, which were also built until 1950. Fred Hawksworth, the last GWR CME, was offered the option of outside Walschaerts valve gear on four out of thirteen alternative schemes for his 'County' 4-6-0, although even in wartime he opted for the traditional and less accessible Stephenson valve gear. It was surely significant that, like his two predecessors in office, Charles Collett and George Churchward, Hawksworth's career had not included any running shed experience. In contrast, Collett's and Churchward's contemporaries, Nigel Gresley and Richard Maunsell, had both had charge of locomotive sheds on the LYR earlier in theirs, and their respective maiden locomotive designs for the GNR in 1912, and the SECR in 1917, both 2-6-0s, featured readily accessible outside Walschaerts valve gear.

Nevertheless, the technical sophistication of GWR No. 98 when it took its bow back in 1903 had in its turn contrasted sharply with that of the other less functional, but arguably more elegant, 4-6-0 express passenger engines that were also introduced during that same year.

Builder's official photograph of Glasgow & South Western Railway 4-6-0 No. 384 (1903). *Rail Archive Stephenson*

Those on the Glasgow & South Western and Great Central also both had outside cylinders, while that on the Caledonian Railway, with inside cylinders, amounted to little more than an enlarged and elongated 4-4-0 (see Chapter 5).

It is an interesting fact that the detailed design, as well as the construction, of both of the GSWR and the GCR 4-6-0s was entrusted to commercial builders, Sharp, Stewart & Co., and Beyer, Peacock & Co. respectively, although a prior detailed specification would appear to have been drafted by the GSWR, which had also considered a large four-cylinder 4-4-0. But by the time the Scottish engines were completed they carried very early North British Locomotive Company works plates bearing five-digit serial numbers. The working drawings handed over with them, as was standard practice, would later be resorted to by Kilmarnock Works when it built seven near repeat 4-6-0s seven years later. Many of the Beyer, Peacock drawings for their part would establish new standards as regards boilers and cylinders in particular, which would later be worked to by the GCR when building its own locomotives in house at Gorton Works 'next door'.

In his two new 4-6-0s on the Caledonian Railway, which were purely railway-designed and built, John McIntosh at St Rollox boldly combined the largest cylinder dimensions (21in by 26in) and highest boiler working pressure (200lb/sq in) encountered in British practice to date, to produce a locomotive that was intended to haul the heaviest trains over Beattock summit without any assistance. Both these dimensions would later be scaled down, but five improved 4-6-0s followed in 1906, which were led by the legendary No. 903 *Cardean*.

By 1906 the first of the rather smaller and black LNWR Experiment 4-6-0s were on the scene, handing over to, or taking over, West Coast expresses from McIntosh's blue

4-6-0s at Carlisle. There were never more than seven of his 6ft 6in 4-6-0s, and only six after the tragic Quintinshill disaster in May 1915, when No. 907 was written off together with a 2-year-old Dunalastair IV 4-4-0. By no means all of this elite band worked between Glasgow and Carlisle at any one time, however, for these 4-6-0s were also variously stationed in Edinburgh and at Perth, from where they worked as far north as Aberdeen. On the other hand, in typical Crewe fashion the LNWR went on to build 105 Whale Experiment 4-6-0s up to 1910, which were quickly followed in 1911 by the first of 245 similar but superheated Bowen Cooke Prince of Wales engines. Many of these were built after 1917, and having 6ft 3in diameter coupled wheels, although officially regarded as express passenger engines, they proved to be very useful mixed traffic locomotives during a very demanding period.

In mid-1909 a locomotive exchange had taken place between the LNWR and the CR, whereby *Cardean* itself ran between Preston and Carlisle, pitted against Experiment No. 2630 *Buffalo*, and employing the new LNW dynamometer car behind their respective tenders to measure drawbar pull and speed. For many years CR No. 903 was credited with having put up an uncharacteristic and quite exceptional performance south of the border, i.e. the development of an estimated 1,400 to 1,500 *equivalent* drawbar horsepower (i.e. corrected for gradient) sustained for sixteen minutes. Only later it became apparent that the calculations had inadvertently assumed the weight of the *train* to be 361 tons, whereas in fact this figure had also included the weight of the locomotive. After deducting this, the result was a respectable but rather less remarkable 925 EDBHP, compared to the corresponding estimated 815 EDBHP developed by the smaller LNWR engine. Comparative coal consumption was 61.1 and 55.7lb/mile. Another member of the Experiment class, No. 1405 *City of Manchester*, was cleared to work between Carlisle and Glasgow, despite being some 7in taller

The legendary Caledonian Railway 4-6-0 No. 903 *Cardean* built at St Rollox Works in 1906, seen in its prime. *Former Ian Allan Archive*

to its chimney top than its CR counterpart, the ill-fated No. 907. Here again the Crewe engine was very slightly more economical at 57.3 as against 58.5lb/mile by CR No. 907. These appear to be the only known records of actual coal consumption by the Caledonian 903 class.

Very different dark blue inside-cylinder 4-6-0s made their first appearance at the very end of 1911, with the emergence from Stratford Works of Great Eastern Railway No. 1500, which represented quite an advance in size and power over the classic 'Claud Hamilton' 4-4-0s introduced

The pioneer Great Eastern Railway 4-6-0 No. 1500 stands resplendent in its ultramarine livery outside Stratford Works, probably in January 1912 when brand new. Only three years later, this magnificent blue livery would rapidly begin to disappear. The small balance weight on the leading coupled wheels, seem here, featured only on the first five engines. *Former Ian Allan Archive*

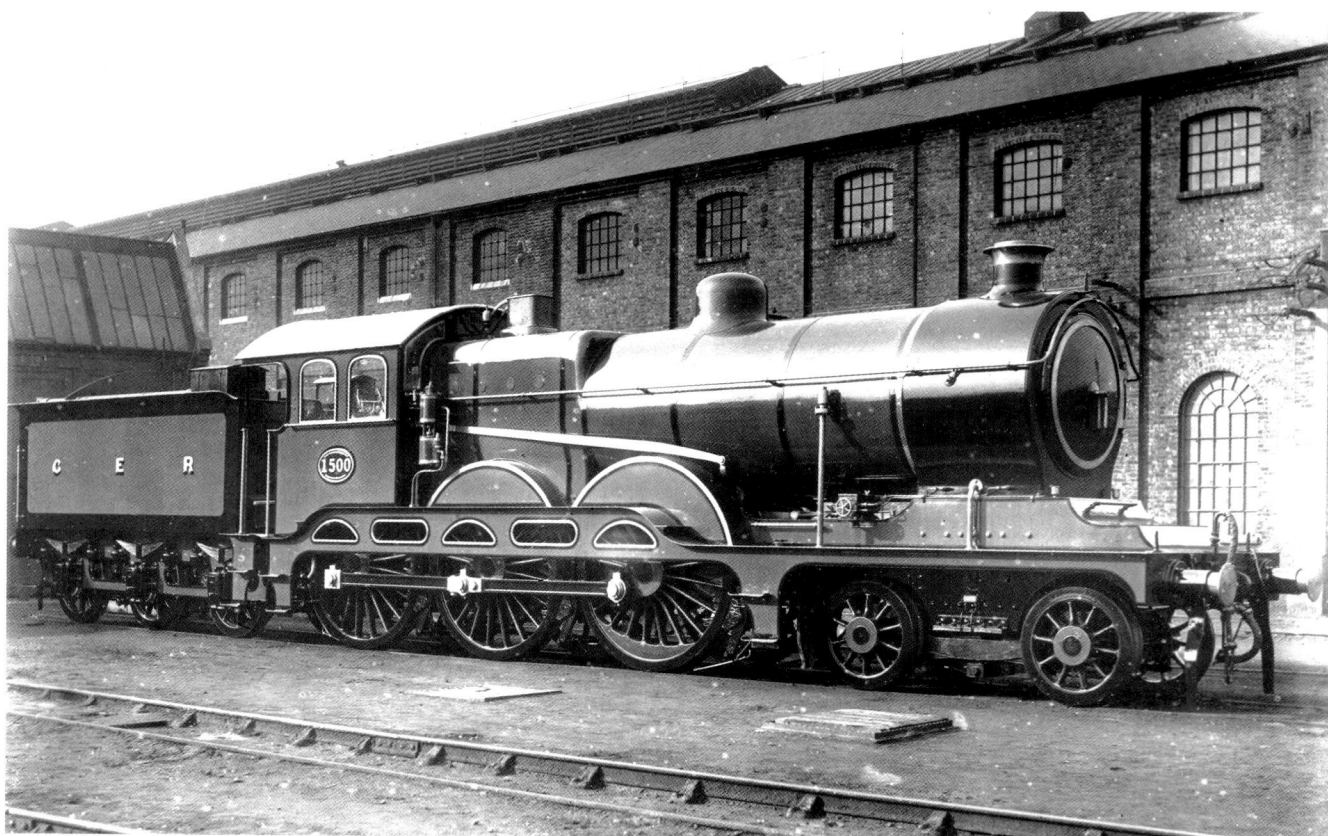

in 1900, and production of which had just ceased after nearly twelve years. Although appearing during the reign of Stephen Holden at Stratford, it had been initiated under his father, James, in February 1908 five months before his retirement. Shortly before its appearance, four of a final batch of ten 'Clauds' had been superheated, and such had been extended to the new 4-6-0. When work on this had begun nearly four years earlier naturally it was to have been non-superheated, and the original cylinder drawing that still survives indicates slide valves, rather than the piston valves that were actually fitted. When compared to many of its contemporaries this 4-6-0 was circumscribed, not only by short turntables, but also by stringent weight restrictions. It was tailored to an axle load of just over 15½ tons, and in total it was only about 4 tons heavier than a Great Central Director 4-4-0. For an inside-cylinder locomotive the long piston stroke of 28in was unusual.

Unlike other British 4-6-0s with inside cylinders, the firebox on the GER engines was *centred* directly above the middle coupled axle, which resulted in an unusually long and commodious cab. Thirty-six 4-6-0s were initially built by the GER between December 1911 and March 1915 to four separate orders. The last of these unusually was for six engines, in order to include a replacement for No. 1506. The latter engine, new in February 1913, had been severely damaged beyond economic repair when in collision with a light engine at Colchester only the following July. Ironically, a fine official portrait of No. 1506, brand new in shop grey livery, was quite coincidentally published in that month's issue of *The Railway Engineer*.

Not apparent in the above mentioned photograph were the very prominent balance weights in the leading coupled wheels that featured on No. 1505 onwards. After the initial batch of five engines (as illustrated), instead of the outside crank pins being set at 90 degrees to the inside crank pins as per usual practice, these were now set so as to directly correspond, as had been the practice of William Stroudley on the LBSCR.[2] Although evidently resulting in considerably reduced wear and tear on the running gear, the very much larger balance weights in the leading coupled wheels would have inflicted a much greater hammer blow, which would have been 6.5 tons at 70mph. This evidently did not prevent several of these engines, on account of their low static axle load, having recently been rendered redundant by the advent of the new Gresley Class B17 three-cylinder 4-6-0s, from being transferred from East Anglia by the LNER in the early 1930s to work on the former Great North of Scotland lines. These had hitherto not seen anything larger than a basic c. 46-ton 4-4-0 design, latterly superheated, that had been in intermittent production between 1893 and 1921.

Although No. 1506's boiler and tender were recovered for further use, this running number was not used again, the replacement locomotive being numbered 1535. A further thirty-five 4-6-0s were built at Stratford and by William Beardmore & Co. between 1915 and 1921, which were later followed by a final ten in 1928 by Beyer, Peacock & Co. in 1928 for the LNER, in order to alleviate a serious short-term motive power shortage on its Great Eastern section, and which unsuccessfully, were equipped with Lentz poppet valve gear. A number of existing engines were fitted for a time with ACFI feed water heaters, in order to try and reduce their coal consumption.

Three-quarters of the class, including the last ten, were ultimately rebuilt with significantly larger 5ft 6in diameter round-topped boilers and long travel valve gear by the LNER between 1932 and 1944, although surprisingly there was no commensurate increase in boiler pressure. Immediately thereafter the gradual withdrawal of the class began, although the new boiler type continued to be built at Stratford Works until as late as 1955. The original 1911 4-6-0 design had only just become extinct in 1954, with the withdrawal of the 42-year former GER No. 1502, which still retained its short travel valves, and which latterly had been running fitted with one of the last Belpaire boilers that had been built in Manchester in 1928.

Just one year after GER No. 1500 was unveiled, Great Central 4-6-0 No. 423 *Sir Sam Fay* made its debut in December 1912. The contrast was considerable, not least in regard to the considerably greater weight of the GC engine, i.e. 75¼ versus 63 tons. Although both had almost the same grate area, the Gorton boiler barrel was 5in greater in diameter and about 5ft longer. It had been intended to send this engine to the 1913 *Exposition universelle et internationale* (World Fair) at Ghent, and so it was given a special finish that even included a brass cap to its chimney, yet three of the remaining five engines in the class that were completed in 1913 were finished in black goods livery. Although this was no doubt for economy reasons, this did not augur well because these six 4-6-0s were very quickly replaced on front-rank passenger duties by the new, smaller Director 4-4-0s, and were indeed demoted to working overnight fitted goods trains also between Manchester and London.

Meanwhile, the period 1905–13 also witnessed something of a spate of four-cylinder 4-6-0 designs that, with the notable exception of the outstanding GWR Stars and Saints already discussed, were not overly successful, and which displayed a remarkable diversity of cylinder layouts and valve gear arrangements.

Although the big inside-cylinder 4-6-0s on the Caledonian Railway were very much in the Drummond tradition, when the great man himself began to build 4-6-0s for the LSWR in 1905, particularly for service between Salisbury and Exeter, these were of elephantine appearance with large 5ft 9in diameter boilers and four cylinders. Not only that, but whereas the outside cylinders had *inverted* Walschaerts valve gear operating slide valves located *beneath* them, the inside cylinders were served by Stephenson valve gear that worked slide valves set between. History has shown that combining two different valve gears,

[2] Skeat, W. O., GER 1500 class 4-6-0 locomotives, *Trans. Newcomen Society*, Vol. 42, 1970, pp. 75–106.

## Table 15 Comparison of GNR large 4-4-2 and LSWR four-cylinder 4-6-0 boilers

| | Boiler diameter | Tube length | Tube HS sq ft | F'box HS " | Grate area | Free gas area |
|---|---|---|---|---|---|---|
| GNR 251 | 5ft 6in | 16ft 0in[1] | 2337 | 141 | 30.9 | 5.51 |
| LSWR F15 | 5ft 9in | 14ft 2in[2] | 2210 | 160 + 357[3] | 31.5 | 4.99 |
| LSWR G14 | 5ft 0in | 14ft 2in[2] | 1580 | 140 + 200[3] | 31.5 | 3.58 |

[1] 2¼in tubes, [2] 1¾in tubes, [3] firebox water tubes.

one having constant (Walschaerts), and the other variable lead characteristics (Stephenson) in one locomotive, the resulting trifling differential of just a small fraction of one inch invariably resulted in poor performance. Despite this fortunately rare impediment, and only 6ft diameter coupled wheels, the first Drummond 4-6-0 nevertheless allegedly attained the remarkably high speed of 82mph when on test![3] Five engines of Class F13 were completed at Nine Elms Works in late 1905 when, then given out as 73 tons, their weight was seemingly precisely the same as that of the Caledonian 903 class a few months later, despite the former having much larger boilers and four cylinders. Three years later the official weight of the F13 was indeed revised up to 76½ tons, with the loading on each of the coupled axles now very substantially correspondingly increased by 1½ tons.

A supposedly improved version of the F13, having the same boiler but larger cylinders, classified Class E14, and now having piston valves worked by outside valve Walschaerts valve gear mounted the right way up, was built in November 1907. Although five locomotives had been authorised, only the first, No. 335, was actually built and this was stated to weigh 74 tons.

Reading between the lines there were possibly weight issues with the civil engineer concerning these distinctly cumbersome 4-6-0s. The G14 4-6-0, with smaller 15in by 26in cylinders, was introduced just four months after the appearance of No. 335 and officially weighed only 71 tons. This lower weight had clearly been achieved by drastically reducing the boiler diameter to only 5ft, and thereby the evaporative heating surface by 30 per cent. The boiler nevertheless still retained a 9ft 6in long shallow firebox with a horizontal 31½ sq ft grate as before, which was likewise set above the intermediate and trailing coupled axles LNW Experiment fashion. This common feature must have made all the Drummond 4-6-0s exceedingly difficult to fire, and yet there were still more of these to come, which were in no way altered in this respect.

The scaled down 4-6-0 boiler adopted the 5ft diameter and precisely the same tube arrangement as the Drummond S11 4-4-0 of 1903. The result of this 'mix and match' surely ranked as one of the worst-proportioned British locomotive boilers ever designed. It made for an interesting contrast with that of the Great Northern large 4-4-2, whose wide firebox afforded a very similar grate area of 31 sq ft, both being unusually large for that period.

The G14 boiler also featured on the essentially very similar P14 4-6-0s, built at Eastleigh in 1910–11, and finally on the ten Class T14 'Paddleboxes' that followed immediately on their tail and into early 1912. These were popularly referred to as such on account of their voluminous and continuous splashers, which enveloped the 6ft 7in diameter coupled wheels. These had been enlarged from the previous 6ft, and it was in order to maintain a similar starting tractive effort that the boiler pressure was correspondingly increased from 175 to 200lb. All four cylinders were now set in line beneath the prominently 'swept-out' smokebox, which enveloped the outside pair. The outside Walschaerts valve gear worked the inside piston valves via rocking levers, which at last made for a much tidier arrangement. These engines were generally regarded as the best of a very poor bunch, and were very soon superheated by Robert Urie, beginning in 1915. After further much-needed alterations to their axlebox lubrication arrangements by the Southern Railway in 1930, they survived until 1948–50, with the exception of one engine that was destroyed in an air raid at Nine Elms in 1940.

On the other hand, quite exceptionally on the part of the Southern Railway, nine of the ten G14/P14 4-6-0s, although they had earlier been scheduled for drastic rebuilding by Robert Urie, were withdrawn all together at a stroke in January 1925. Then they were less than seventeen years old, although the tenders were usefully passed on to other locomotives. In contrast, the solitary E14 had been rebuilt into the new Urie H15 class after only seven years in

*Opposite top:* London & South Western Railway Class F13 four-cylinder 4-6-0 (1905). *Former Ian Allan Archive*

*Middle:* LSWR Class G14 4-6-0 (1908) with reduced-diameter boiler. *Former Ian Allan Archive*

*Bottom:* LSWR Class T14 ('Paddlebox') 4-6-0 (1911) No. 447. *Former Ian Allan Archive*

The large casing on the outside of the firebox of each of these locomotives was associated with the cross water tubes contained within. This troublesome feature of many of Dugald Drummond's designs was hastily removed following his death. Also it was not perpetuated by Peter Drummond after he transferred to the GSWR in 1912.

[3] Bradley, D. L., *Locomotives of the London & South Western Railway*, Part 2, The Railway Correspondence & Travel Society, 1967, p. 156.

1914. It remarkably lasted in this form until 1959, while its likewise little-used F13 predecessors were similarly completely rebuilt in 1924, and also remained in traffic until 1955–61. In each case only the original boiler barrels, and tenders were retained, as were the 6in longer shallow fireboxes with level grate of the original Drummond design, although with the omission of his characteristic cross water tubes. Thus against the odds, on paper at least, old LSWR No. 331 at fifty-six years boasted the longest normal service life of any British 4-6-0. The nearest contestant was rebuilt GWR 'Star' No. 4037 (1910–63).

During the autumn of 1912 design work had begun at Eastleigh on yet another Drummond four-cylinder 4-6-0, Class K15. With coupled wheels of uncertain diameter, this would have reverted to the larger 5ft 9in diameter boiler of the early Drummond 4-6-0s, but in conjunction with the recent T14-style cylinder layout. Owing to Drummond's sudden death this failed to materialise, but a rather similar 4-6-0 with 6ft 6in wheels, 16in by 26in cylinders with 10in diameter piston valves, and a large boiler of similar proportions was later proposed by his brother. In January 1912 Peter Drummond had transferred from the Highland Railway to the Glasgow & South Western, and after he had introduced his exceptionally heavy non-superheated 4-4-0s and 0-6-0s thereon in 1913, he began to consider building a large *superheated* four-cylinder 4-6-0, which unsurprisingly would have owed much to his late brother's recent K15 proposal on the LSWR. However, Kilmarnock progressed no further than producing an initial diagram for this in April 1914, made by the draughtsman David Smith, who had followed Drummond from Inverness. The almost horizontal grate area would have been no less than 33¼ sq ft; no estimated weights appear to have been calculated.[4, 5]

The utterly haphazard manner in which 4-6-0 design initially progressed on the London & South Western Railway was in sharp contrast to the careful design, albeit with disappointing results, of the Hughes four-cylinder 4-6-0s on the Lancashire & Yorkshire Railway. On their appearance in 1908 they became known as 'Dreadnoughts', a term that reflected the growing arms race with imperial Germany in terms of battleship construction that dominated the news at that period. Twenty of these 4-6-0s had been authorised in February 1907, only three months after the emergence of the LYR's first locomotive to be fitted with a superheater, 0-6-0 No. 898, and hence too early to attempt to incorporate such a major innovation in the new engines. These were therefore designed with slide valves that were worked by internal Joy valve gear, with rocking levers to operate the valves above the outside cylinders. By the standards of the period, the boilers were quite well proportioned, with a generous diameter of 5ft 9in, in conjunction with a grate area of 27 sq ft, as in the GWR Saints and Stars. Particular

attention was paid to the balancing, and whereas the piston stroke was the usual 26in, the *throw* of the coupling rods was only 10in.

After the first engines entered traffic some alterations were found to be necessary to the draughting arrangements in the smokebox, in order to improve steaming, while ducting was soon fitted to the underside of the ashpan, below its hump to clear the trailing coupled axle, in order to augment the air supply to the rear portion of the firegrate. (It is not clear how readily the rear wheelset could then be dropped on shed, should the axleboxes require attention.)[6]

In addition to workings west of the Pennines, which were centred on Manchester and Liverpool, like the Aspinall 4-4-2s before them, the 4-6-0s' duties would take them to York, to work the heavy 450-ton NER/LYR joint Newcastle–Liverpool expresses, and also to Hull. These turns involved working over North Eastern metals beyond Altofts Junction, just east of Normanton, and Goole respectively. Coincident with the imminent emergence of the lead engine, No. 1508, in June 1908, a weight diagram was forwarded to the NER civil engineer at York. This, however, gave only the preliminary *estimated* weights, which had been 72½ tons for the engine, with an axle load of 18½ tons. On completion it became apparent that the *actual* weights were somewhat higher at 77 tons and 19.8 tons respectively. This is possibly why, on presumably discovering the true figures, the NER then barred the engines from working into Hull, nor were they permitted to enter Leeds until 1919. The LYR 4-6-0s also worked south over the Great Central from Penistone into Sheffield Victoria.

As early as 1911 the six 4-6-0s allocated to the LYR's No. 1 shed, Newton Heath, were displaced from their duties by new superheated 2-4-2 tank engines! As early as 1909 one 4-6-0 had been fitted with a form of smokebox superheater that did not impede access to the fire tubes, while in late 1913 No. 1509 was fitted with a Field tube superheater. The outbreak of war undoubtedly delayed the application of conventional superheating to the class, but in 1916 drawings were prepared for new boilers with twenty-eight element superheaters, which were based on those that were already established on the large 0-8-0s. Five of these boilers were ordered in January 1918, but for economy reasons they were completed *without* superheaters, which were then put onto the five engines whose slide valve cylinders remained in the best condition. This explains why these five were later scrapped during 1925–26 without having been rebuilt as were the remaining fifteen during 1920–21.

By 1918 the LYR 4-6-0s were in very poor condition and their coal consumption was prodigious, even approaching 100lb/mile in extreme instances according

---

[4] Smith, D. L., *Locomotives of the Glasgow & South Western Railway*, David & Charles, 1976, pp. 142–4. David Smith (draughtsman) was not David L. Smith, although the two men were acquainted with each other.

[5] Barnes, R., *Locomotives That Never Were*, Jane's, 1985, p. 29, artist's impression of proposed P. Drummond GSWR four-cylinder 4-6-0.

[6] Hughes, G., Locomotives designed and built at Horwich, *Proceedings of the Institution of Mechanical Engineers*, 1909, pp. 561–639.

to Eric Mason. Many by this time were laid up awaiting conversion to superheating, which would also require new cylinders incorporating piston valves, for which the associated Walschaerts valve gear was not fully designed until 1919. For some reason Horwich fought shy of going for long-lap/long-travel valves, despite having apparently successfully applied these a full decade earlier on a four 4-4-0 rebuilds. The 4-6-0 No. 1522 was the first 'rebuild', completed in October 1920, although having been given a new boiler, new frames, new cylinders and even new coupled wheel centres, it was in truth effectively a new engine. Indeed, even fresh number plates were cast that

now bore the renewal date. This was certainly a considerable improvement on the original 1908 design, *relatively* speaking, reducing their coal consumption from a roughly assessed 7lb/DBHP hour in 1918 down to 4lb. In more practical everyday terms, this amounted to a still high 60lb per mile, which in the 1930s would tell heavily against them, resulting in many engines suffering a very premature demise. The Hughes superheated 4-6-0s did at least bequeath their basic cylinder and valve gear layout to the Stanier LMSR Coronation 4-6-2 (1937), and they also had the distinction of never having been involved in a serious railway accident.

LYR 4-6-0 No. 1515 leaves Poulton-le-Fylde with an up express from Blackpool. *J. M. Tomlinson/Rail Archive Stephenson*

A 1921-built LNWR 'Claughton' 4-6-0, No. 1097 *Private W. Wood V.C.*, one of three named after company employees who gained the supreme award for their gallantry during the recent war, in post-1914 plain black livery at Manchester, London Road Station in 1923. *T. G. Hepburn/Rail Archive* Stephenson

Thirty-five superheated 4-6-0s were then built new at Horwich between August 1921 and August 1923, after which twenty more, which had originally been ordered as 4-6-4 tank versions, were delivered in 1924–25. There was a later proposal in 1929 to rebuild these with new 200lb pressure boilers similar to those being fitted to the ex-LNW Claughtons, and they were also included in an ambitious January 1933 plan to rebuild numerous pre-1923 locomotive classes with highly standardised Stanier taper boilers. This proposal was almost immediately abandoned in favour of a ruthless 'scrap and (new) build' policy. Indeed, scrapping of the Hughes superheated 4-6-0s thereupon began as soon as late 1933 and was originally scheduled to be completed by 1939 as replacement fireboxes routinely became due. However, the tense international situation reprieved several, including one so-called 'rebuild', former LYR No. 1516, which managed to survive until February 1949, by which time it was theoretically and quite exceptionally forty years old.

On the LNWR the year 1909 had not only witnessed *Cardean* working north of Crewe, but also a GNR large-boilered 4-4-2 operating south of Crewe, while a Whale Precursor 4-4-0 plied between King's Cross and Leeds. More significant could and should have been the appearance of non-superheated GWR Star No. 4005 *Polar Star*, which also worked between Euston and Crewe in August 1910, while

Experiment No. 1471 *Worcestershire* plied between Paddington and Exeter. However, these latter events had not been at the LNWR's behest at all, but were by way of a rain check by the GWR following criticism by its general manager of the allegedly high cost of Churchward's locomotives. That said, the contemporary cost of LNWR locomotives does not seem ever to have been revealed.

In October 1911, the same month that Crewe Works produced its first superheated 4-6-0, No. 819 *Prince of Wales*, it also received an order for ten superheated four-cylinder 4-6-0s of new design, which on past reckoning should all have been in traffic within twelve months. However, *The Locomotive* for February 1912 noted:

> 'The new 4-cylinder simple engines will not be out probably before the autumn of this year; the work which was already in hand having been put on one side for the present.'

The reasons for this hiatus remain unknown. All four cylinders were to be in line abreast driving onto the leading coupled axle, and with the novelty (for the LNWR) of having outside short lap/short travel Walschaerts valve gear. When these finally appeared the contemporary observer could have been forgiven for assuming that they had been inspired by the recent and very similarly configured Prussian State Railway (KPEV) four-cylinder simple Class S10 4-6-0, of which a total 202 was built between 1910 and 1914. However, according to Kenneth Cantlie the inspiration had in fact been another, rather older and much smaller German 4-6-0, a *compound* mixed traffic engine on the Bavarian State Railway (Class P3) that had been introduced in 1901.

No early schemes for the LNW four-cylinder 4-6-0 have since come to light. Cantlie (who would later go on to design the notable British-built 4-8-4s for China) recalled that when he was a premium apprentice at Crewe during 1916–20 he was shown one such by Tommy Sackfield, its long-standing locomotive leading draughtsman. Cantlie had remarked that this showed a shorter chimney than that on the engine as built. Sackfield replied that it had been intended to fit a larger diameter boiler, but that the civil engineer, Ernest Trench, had objected to the estimated weights being too high. Although the CME, Bowen Cooke, pointed out to him that as the perfectly symmetrical four-cylinder arrangement therefore required no reciprocating balance weights, hammer blow would be eliminated entirely, so that a higher axle load should therefore be permissible. Unfortunately, Trench would not be moved.[7] The boiler diameter was therefore reduced to the well-established standard 5ft 2in, although the firebox, possibly already designed to match the original proposed dimension (about 5ft 5in), was nevertheless retained.

Compared with the very long maximum valve travel of the Stars, on the Claughton this was to be only 4¼in in conjunction with 1in lap, albeit in conjunction with short and direct steam ports. Illogically, in a distinctly retrograde fashion this compared unfavourably with the 5½in travel and 1¼in lap on the George the Fifth 4-4-0s. This resulted in a restrictive maximum port opening of only 1 in as against 1½in on the 4-4-0, and almost 1¾in on the GWR Star. As the Crewe pioneers of hot steam, the design of the Georges would, no doubt, have been subject to some direct advice from Dr Schmidt, which unfortunately was not further fully heeded.

In November 1912 *The Locomotive* informed its readers about the LNWR:

'The first of the long expected passenger engines will shortly be put in hand, one will be named after the Chairman of the company. The series will be officially known as the "Claughton" class.'

In fact, LNWR No. 2222 *Sir Gilbert Claughton* emerged from Crewe Works only two months later in mid-January 1913, when it made an initial trial trip, nameless and in grey primer, to Manchester. The engine was duly inspected at Crewe two months later by the Locomotive Committee, including Sir Gilbert himself, who had been a former Beyer, Peacock & Co. apprentice. On that occasion an indicator shelter was attached, but no test data derived from No. 2222 appears to have survived. However, late in 1913 full power trials with the ninth engine, No. 1159 *Ralph Brocklebank*, new in July, were undertaken. On 2 November, with 435 tons behind the tender, No. 1159 ran north from Euston to Crewe. Two days later, with 360 tons, it departed Crewe 7¼ minutes late yet

arrived at Carlisle 9¼ minutes early! While regrettably no reference was made to coal and water consumption, the tests were otherwise reported in some detail in *The Railway Gazette* for 1 May 1914:

### Table 16 Full power trials with LNWR 4-6-0 No. 1159 in November 1913

|  | London–Crewe | Crewe–Carlisle |
|---|---|---|
| Distance | 158 miles | 141 miles |
| Train weight (gross) | 435 tons | 360 tons |
| Overall time | 2 hr 39 min | 2 hr 20½ min |
| Average speed | 59.6mph | 59.4mph |
| Maximum indicated horsepower | 1,548 @ 66.5mph | 1,669 @ 69.0mph |
| Average ditto | 1,358 | 1,387 |
| Average drawbar horsepower | 975 | 920 |

The peak IHP of 1,669, which was registered while passing through Tebay, was remarkably high and unsurpassed in Britain for many years. Unfortunately no corresponding values for the GWR Stars are on record, although in the *Journal of the Stephenson Locomotive Society* for January 1970, Arthur Cook suggested a modest maximum of 1,345 for a superheated and 1,250 IHP for a non-superheated Star. When tested in 1926 the new and larger Southern Railway four-cylinder 4-6-0 No. 850 *Lord Nelson* for its part delivered a maximum of merely 1,425 IHP at 55mph. On the basis of the above, even when considered in relation to its 30½ sq ft grate area, which fell midway between those of the Swindon and Eastleigh designs, the power output developed by the Crewe engine was truly remarkable, particularly with regard to its inferior boiler and cylinder design.

Although No. 1159's performance had apparently been quite outstanding during these tests, in ordinary everyday service the work of the new Claughtons proved to be distinctly variable. The long grate proved difficult to fire, with the firebed wearing thin at the point where the forward-sloping section levelled out above the trailing coupled axle, which itself could suffer from serious lubrication problems. Steaming qualities were not good, despite the commendably generous grate area. In order to obtain a commensurate evaporative heating surface it seems possible that the original proposed tubing arrangement had been shoehorned into the smaller diameter boiler. Indeed, if a direct comparison is made with the Prince of Wales 4-6-0 boiler having the same 5ft 2in diameter, on the Claughton the tube and flue bundle was packed about 10 per cent more densely. This inevitably resulted in reduced water spaces, which would have been to the detriment of satisfactory water circulation and consequently steaming capacity. Conditions would undoubtedly have been much improved by a modest 3in increase in boiler diameter. It is therefore interesting to read in *The Locomotive* for February 1914:

'It is reported that ten new "Claughtons" which are on order will have *slightly larger boilers* than those now running. These engines, however, will not be put in hand just yet, nor is it anticipated that they will be ready for the coming summer traffic.'

Again no documentation now exists in support of this assertion, and although this batch did not in fact appear until 1916, they were nevertheless fitted with the same standard boiler as before. Although the Claughtons had been designed with 16in diameter cylinders, these soon became reduced to 15¾in as the standard. Bowen Cooke had been well aware of the shortcomings of the Claughtons, of which only a relative few had been turned out by August 1914, after which any significant development work on these was put on hold for the duration of the war. Despite the consequent government embargo on building new passenger locomotives, thirty Claughtons were approved by the Ministry of Munitions to be built in 1917 to work the essential Naval (personnel) Specials between London and Carlisle. In October 1918 a modified tubing arrangement was proposed with 105 2¼in tubes in place of 149 (originally 159) 1⅞in tubes, but this was not proceeded with.

When post-war construction resumed in 1920, the firegrates on the new engines were given a continuous slope. Kenneth Cantlie also suggested reviving an old Francis Webb patent whereby the smokebox would be equally divided by a horizontal partition, with the upper and lower portions each having its own separate blastpipe and chimney, as had briefly been tried on two Webb 4-4-0s and a 2-4-0 around 1897.[7] However, around this time, in October 1920, Bowen Cooke died suddenly while still in harness, while Cantlie had departed from Crewe to work overseas. No significant modifications were in fact made to the Claughtons before 1923, although there had been an experimental application of a single thermic syphon inside the firebox of one engine, which by all accounts leaked like a sieve.

In the 1928 Bridge Stress Committee data enabled the Claughton to be cleared to run on the LMSR Midland Division, subject to modifications to cabs and boiler mountings in order to comply with the differing former Midland loading gauge. With its 19¾-ton axle load the Claughton and its perfect balancing, the hammer blow at any speed was nil, whereas on the smaller Prince of Wales, whose axle load was only 18¼ tons, at 5rps (67mph) the hammer blow was 7.9 tons. (On the George the Fifth 4-4-0s, in fact, this was no less than 9.8 tons!) At a common line speed of 60mph the combined impact on the track and underlying structures made by the Prince would actually have been 25 per cent, or 5 tons greater than that of the significantly larger Claughton. Quite

unwittingly, some 'Princes' had already been permitted to operate out of St Pancras several years earlier!

Also in 1928, the LMSR rebuilt at random twenty Claughtons of varying age with new 5ft 5in diameter boilers pressed to 200lb, additionally equipping ten of these with Caprotti poppet valve gear (of which five were altered to Harry Holcroft's proposed 135 degree crank setting for four-cylinder locomotives). This same boiler type was employed when forty-two further engines were officially very heavily renewed as three-cylinder Patriots between 1930–33, while the remaining unaltered engines were rapidly scrapped during 1932–35, some when only twelve years old. Withdrawal of the twenty re-boilered four-cylinder engines even began as early as 1935, although the last survivor, built in 1921 and reprieved by the war, enjoyed a final eight years of service in splendid isolation, and it was still attired in LMSR red livery when it finally succumbed in 1949.

## 4-6-0s that never were

By 1912, of the fourteen major railway companies, only the Midland and South Eastern & Chatham had not progressed beyond the 4-4-0 to the 4-4-2 and/or the 4-6-0 to work their heaviest express passenger services. This was not for want of trying, however, at least on Ashford's part. In 1906 Harry Wainwright had proposed otherwise similar outside-cylinder 4-4-2 and 4-6-0 locomotives, which were followed in 1907 by an inside-cylinder 4-6-0 that would have been very much in the style of his new Belpaire Class E 4-4-0.[9] As on the Midland, weight restrictions on the SECR bedevilled such adventures, sometimes on account of a single under bridge on a particular route, and later a larger superheated outside-cylinder 4-6-0 that was proposed in 1912 duly fared no better. At Derby a rather ungainly four-cylinder compound 4-6-0 had been sketched out in November 1907, but the Midland Railway's policy in any case was to work frequent light expresses that were within the capabilities of a 4-4-0, and no further attempts whatever were made in this particular direction before 1923.[10] Reference has already been made in Chapter 6 to the North British Railway's 1907 4-6-0 proposal.

Interestingly, there is no indication that the Great Northern at Doncaster at any time considered building a 4-6-0 as opposed to a 4-4-2. However, in 1915 a four-cylinder 4-6-2 in two alternative forms was sketched out, having either a narrow or a wide firebox in conjunction with a 5ft 6in diameter boiler as on the large 4-4-2s. A little earlier, in October 1913 the Caledonian Railway at St Rollox, which never seemed to tire of designing yet another 4-6-0, had mooted a tentative four-cylinder 4-6-2, with similar alternative firebox arrangements, likewise in outline only. It is likely that,

[7] K. Cantlie to author, August 1978.

[8] F. W. Webb, BP 29,240/1896 (Divided smokebox with separate blastpipes and chimneys).

[9] Bradley, D. L., *The Locomotive History of the South Eastern & Chatham Railway*, Railway Correspondence & Travel Society, 2nd edition 1980, pp. 134–45.

[10] Radford, J. B., *Derby Works and Midland Locomotives*, Ian Allan, 1971, p. 139.

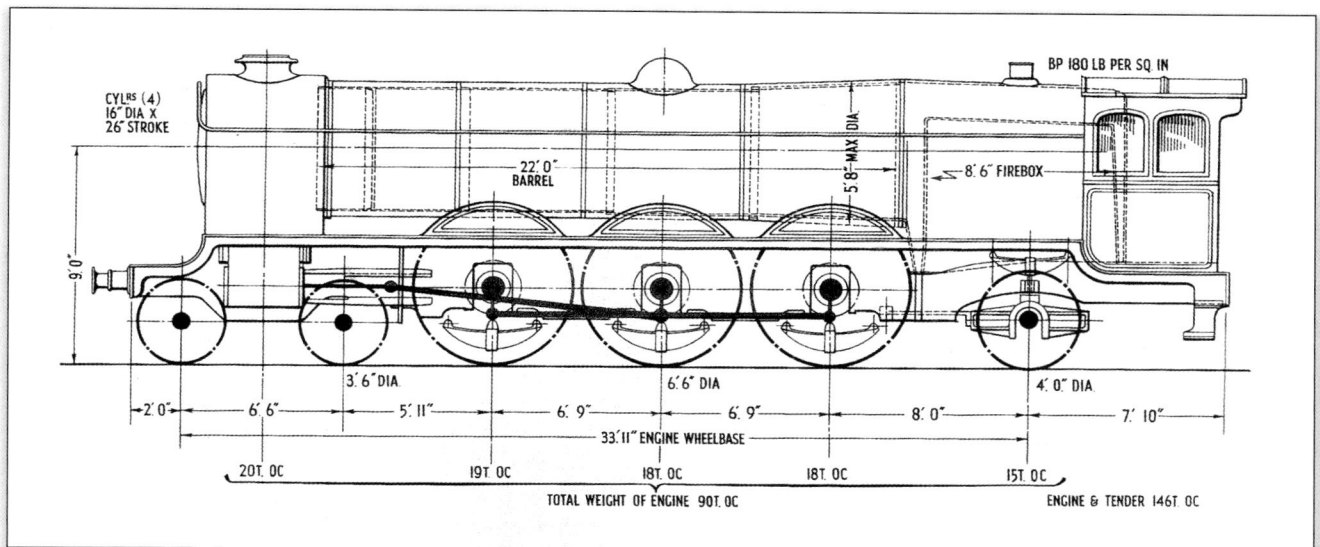

Diagram of proposed CR four-cylinder 4-6-2 (wide firebox version), 1913.

following the recent appearance of the new Claughtons on the LNWR this had been prompted by top management in order to keep abreast of, if not one ahead of the Caledonian Railway's English partner, as John McIntosh had managed to do since 1903. As with *The Great Bear*, however, the back end of the St Rollox 4-6-2 would have left much to be desired, in combination with similarly unduly long fire tubes.

This historic multiple diagram, referred to on p. 65, was prepared at Swindon in January 1901, and laid the foundations for the early Churchward GWR standard two-cylinder locomotives, and ultimately for a great many more. It was made by George Burrows, who would later be promoted to become chief draughtsman. Ably assisted by Oscar Deverell, who designed the boiler for *The Great Bear*, Burrows was therefore very closely involved in developing Churchward's locomotive designs. Sadly, both men would later die in the years which closely followed Churchward's retirement in 1922, when each was aged only in his early fifties.

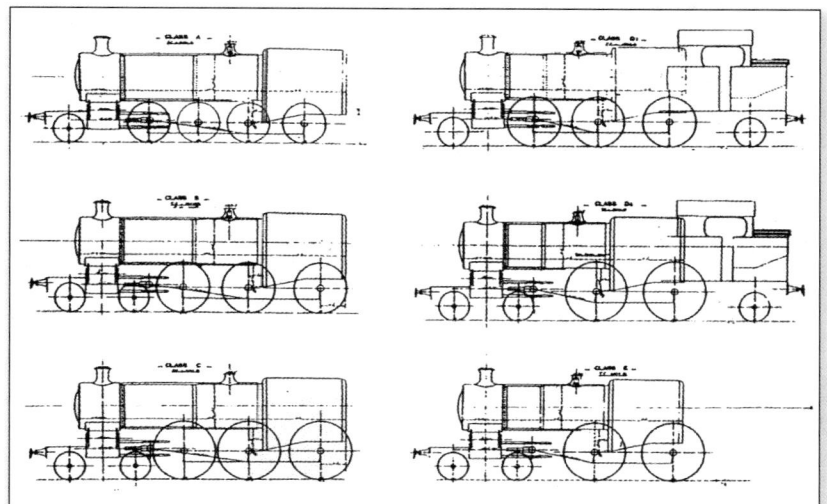

# 8

# The Rise of the
# Heavy Goods Engine

Total goods train mileage peaked at 175 million in 1900, but had declined by 20 per cent to 135 million by 1913. Goods train mileage fell by 10 per cent between 1901 and 1911, although total tonnage carried had meanwhile increased from 411 to 571 million tons. However, as total merchandise tonnage had slightly declined by 6 million tons, the big increase was entirely due to minerals, particularly to service British manufacturing industry, then at its undeniable zenith, and coal for export to Europe and beyond. British coal output had been steadily rising, but owing to international events would actually peak in 1913 at 287 million tons, of which 73 million tons was exported overseas. Coal was lifted in six principal regions:

| | |
|---|---|
| Northumberland & Durham | 58.7m tons (23.0m tons exported) |
| South Wales | 56.8m tons (34.9m tons exported) |
| Yorkshire | 43.7m tons (not recorded) |
| Scotland | 42.5m tons (10.4m tons exported) |
| East Midlands | 33.7m tons (not recorded) |
| North-west England | 28.1m tons (not recorded) |

Coal and coal products, e.g. coke, greatly dominated the mineral tonnage statistics, which also included iron ore and limestone for instance, and which were simply collectively tabulated thus up to 1912. A partial explanation of the apparent paradoxes indicated above, i.e. mileage *versus* tonnage, must surely have been the rapid adoption and construction of powerful eight-coupled goods engines by several of the larger railway companies during the Edwardian period together with greater efficiency. On the LNWR for example, its total number of 0-6-0s had peaked at 1,370 in 1892, when its first 0-8-0 was also built, but this had fallen to only 767 by December 1913, this numerical deficit having effectively been made good by the construction in the meantime of 412 0-8-0s (of which thirty-six Class B four-cylinder compounds had subsequently been rebuilt as 2-8-0s), together with 200 fast goods 4-6-0s that had also entered service on the LNWR after 1902.

In some instances the increase between 1900 and 1913 in mineral tonnage hauled was dramatic, actually doubling on the GCR and GNR. Of the top ten coal haulers listed below, which included the Taff Vale Railway in South Wales, only the Caledonian showed no increase at all, actually registering a very slight decrease. In 1913 at 519 the NBR operated significantly more 0-6-0s than did the CR at 346, which nonetheless also boasted eight 0-8-0s of limited actual utility.

## Table 17 Mineral tonnage (millions) hauled by ten railways in 1900 and 1913

| Railway | 1900 | 1913 | % increase |
|---|---|---|---|
| NER | 41.3 | 59.7 | 45 |
| GWR | 30.0 | 47.6 | 59 |
| LNWR | 34.8 | 44.4 | 28 |
| MR | 23.9 | 43.7 | 83 |
| GCR | 13.6 | 28.8 | 111 |
| NBR | 18.4 | 26.4 | 43 |
| CR | 22.4 | 22.2 | -1 |
| TVR | 15.8 | 19.9 | 26 |
| LYR | 15.4 | 19.5 | 27 |
| GNR | 9.5 | 18.9 | 99 |

The *eight-coupled* heavy goods engine made its first appearance in Britain as early as 1889, almost by accident, when the newly established Barry Railway in South Wales acquired a pair of outside-cylinder 0-8-0s from Sharp, Stewart & Co. which had been left on that builder's hands following the bankruptcy of the Swedish–Norwegian iron ore railway. Essentially of the builder's standard design, the Barry later purchased two more of these 0-8-0s in 1897, which had been built ten years earlier. These had seen service in Scandinavia before being retrieved after much litigation on the part of the builders, which in the meantime had removed from Manchester to Glasgow.

*Above:* The ground-breaking prototype Great Western Railway 2-8-0 No. 97, as originally completed in 1903 with lower-pitched, half-cone, non-superheated boiler. Later renumbered 2800 in 1912, this engine, originally conceived in the first days of the 20th century, was not retired until 1958.

*Below:* The classic Churchward GWR 2-8-0 heavy mineral engine as perfected by 1912, with full-cone superheated taper boiler having top feed, as exemplified by No. 2840 built in that year. These changes can readily be detected with reference to the rare photograph of the prototype No. 97 taken when new in 1903, above. *Former Ian Allan Archive*

By 1897 the LNWR also had built thirty-odd 0-8-0s. The first, turned out in 1892, had been a straightforward inside-cylinder engine, a logical extension of the preceding 500 Webb coal 0-6-0s. This engine, which would not be withdrawn until 1949, albeit in heavily rebuilt form, would later prove to have been the unwitting progenitor of no fewer than 571 more 0-8-0s that would all be built at Crewe over the course of the next thirty years. These were characterised by the employment of Joy valve gear, first adopted by Francis Webb on the LNWR in 1880 on his 18in goods 0-6-0s. By December 1900 the Company had 110 three-cylinder compound 0-8-0s (Class A) in traffic, while the following year it produced the first of 170 four-cylinder 0-8-0s (Class B). It had originally been intended to build more of these,

but twenty were instead turned out, seemingly almost as an afterthought, as fast goods 4-6-0s, the so-called 'Bill Baileys'. These had many major parts in common and with characteristic Crewe thrift allegedly utilised the coupled wheel centres recovered from recently withdrawn Ramsbottom DX 0-6-0 goods engines.

The *fin de siécle* British locomotive building frenzy of 1899–1900 clearly concentrated minds in several quarters regarding the more efficient bulk movement of goods, *mineral* traffic in particular, in the future. By the end of 1903 six British railway companies other than the LNWR had between them put more than two hundred 0-8-0s (plus one very significant 2-8-0) into traffic, in some cases in association with new air-braked American-style, high-capacity (30-ton) bogie coal wagons.

## Table 18 Eight-coupled tender locomotive construction by British railway companies, 1901–14

| Year | LNWR | LYR | GNR | GCR | NER | CR | GWR | Other | Total |
|---|---|---|---|---|---|---|---|---|---|
| Pre-1901 | 111 | 20 | | | | | | 4 | 135 |
| 1901 | 20 | 20 | 1 | | 10 | 2 | | | 53 |
| 1902 | 60 | 18 | 14 | 3 | 13 | | | | 108 |
| 1903 | 60 | 22 | 16 | 15 | 19 | 6 | 1 | | 139 |
| 1904 | 30 | 15 | 9 | 18 | 8 | | | | 80 |
| 1905 | | 5 | | 5 | | | 20 | | 30 |
| 1906 | | 8 | 4 | | | | | | 12 |
| 1907 | | 30 | 1 | 13 | 14 | | 10 | 15 | 83 |
| 1908 | | 2 | | | 6 | | | | 8 |
| 1909 | | | 5, 5* | 15 | | | | | 25 |
| 1910 | 60 | 16 | | 17 | | | | | 93 |
| 1911 | | 4 | | 3, 9* | 20 | | 5* | | 41 |
| 1912 | 46* | 3* | | 89* | | | 14* | | 152 |
| 1913 | 24* | 34* | 3* | 13* | 30* | | 6* | | 110 |
| 1914 | 30* | 3* | 2* | 14* | | | | 6* | 55 |
| Total | 441 | 200 | 60 | 214 | 120 | 8 | 56 | 25 | 1124 |

* Superheated

First off the mark after the LNWR had been the Lancashire & Yorkshire Railway in March 1900, with a neat inside-cylinder 0-8-0 that followed on in its fundamentals from the recent express passenger 4-4-2s. The 5ft diameter boiler in particular was very similar but with a slightly smaller firebox. Talking of which, No. 676 when only seven months old suffered a serious boiler explosion at Knottingley in March 1901, sadly with fatal consequences for the engine crew. The cause was later discovered to have been defective firebox stays. Although the construction of similar engines still continued, during 1903 twenty 0-8-0s were built with stayless Cornish-style corrugated fireboxes. However, these were not entirely satisfactory and all were replaced by conventional boilers between 1912 and 1914. Although many standard 'small-boilered' 0-8-0s were later rebuilt with larger diameter boilers from as early as 1911, one of these somehow retained its original small-pattern boiler up to its withdrawal as late as 1950. On the other hand, the ten engines built as four-cylinder compounds in 1907 would feature among the first LYR 0-8-0s to be retired by the LMSR only twenty years later.

New LYR 0-8-0s from 1910 had much larger 5ft 9in diameter boilers, although their weight, now increased by 9 tons, was disproportionately high, suggesting that some ballasting had also been incorporated in order to boost their adhesive weight. Then from 1912 superheaters were incorporated, although there was a later brief reversion to saturated steam for economy reasons in twenty engines that were built in 1917–18, of which curiously none were ever superheated thereafter. A final forty-three engines, once again equipped with superheaters, were turned out

during 1919–20 to bring the LYR's 0-8-0 grand total up to 295, all of which, like the LNWR 0-8-0s were fitted with Joy valve gear.

Next came the Great Northern Railway, on which Henry Ivatt turned out a prototype inside-cylinder 0-8-0 from Doncaster Works in February 1901. Surprisingly it was not the first eight-coupled locomotive on the GNR, which had purchased two 0-8-0 tank engines from the Avonside Engine Co. thirty-five years earlier, in 1866, to work heavy goods trains across London to the London, Chatham & Dover Railway. Based on Britain's very first eight-coupled engines built for the Vale of Neath Railway, these were modified to the requirements of Archibald Sturrock, the first locomotive superintendent of the GNR (during 1850–66). In December 1903, at the age of 87, Sturrock visited Doncaster Works, which he had established fifty years earlier, and a remarkable photograph was taken of him with Henry Ivatt standing by 0-8-0 No. 405. Thirty nine more 0-8-0s were built during 1902–04, five in 1906–07, and a final ten in 1909, the last five of which were superheated from new following recent trials with the Schmidt superheater on an older 1903-built engine.

Several months later in July 1901 two inside-cylinder 0-8-0s were completed by the Caledonian Railway at St Rollox Works. These were distinguished by an unusually lengthy coupled wheelbase of 22ft 4in, arranged so as to provide adequate space at the front end for the driving machinery, and the same 8ft 6in spacing at the back end in order to accommodate a standard (4-4-0) deep firebox. The inside cylinders had the unusually large diameter of 21in, and so had their slide valves necessarily

Typical Lancashire & Yorkshire Railway superheated large-boilered 0-8-0 with eight-wheeled tender. (No. 1619 was built in August 1919 and withdrawn as early as February 1935.) *Former Ian Allan Archive*

mounted directly above, rather than in between as hitherto on the 0-6-0s.

The first two CR 0-8-0s had been ordered in late 1900 to operate in association with fifty new air-braked 30-ton bogie coal wagons. Five hundred more similar wagons were authorised a year later specifically to handle locomotive coal, together with a further six 0-8-0s. In truth these engines proved to be a doubtful investment as they could haul longer mineral trains than could actually be accommodated in the existing sidings and refuge loops, in association with which the standard McIntosh 812 0-6-0s were quite sufficient, and would remain so up to 1922. Despite this limitation, quite some years later McIntosh's successor, William Pickersgill was asked by the CR directors to provide more powerful eight-coupled goods engines, only to be told that this was not feasible until the necessary infrastructure was put in place at some considerable cost by the civil engineer. It was rumoured that Pickersgill had taken umbrage over this issue, and had even considered tendering his resignation. That said, the Caledonian Railway did give trial to an LNWR G1 0-8-0 in Lanarkshire during the winter of 1919–20, and between 1919 and 1921 it hired more than fifty ex-ROD 2-8-0s of Robinson GCR design from the government, at least some of which briefly carried the CR insignia on their tenders.

Photographs of the Caledonian 0-8-0s are extremely scarce and invariably show them stationary, usually on shed. None appear to exist of one in CR livery actually at work, although it is very surprising that No. 600 was not officially photographed at the head of a train of the new

bogie coal wagons for publicity purposes in 1901. Around 1914 several members of this class were given new main frames that were made noticeably deeper over the second and third coupled axles.

Immediately after the Caledonian Railway, the North Eastern Railway turned out its own first 0-8-0, this time with outside cylinders, from Gateshead Works in August 1901. In several respects this was not so far removed from the Barry 0-8-0s, but it also had a close affinity with the recent S class passenger 4-6-0s with their 4ft 9in diameter boilers and 20in by 26in cylinders. Ultimately a grand total of ninety 0-8-0s was built at Gateshead and Darlington up to 1911, some with slide valves (Class T) and others with piston valves (Class T1). None would ever be superheated, despite new boilers being built for the class even until as late as 1945. The first fifty, all of which were built at Gateshead, were initially finished in the full glory of the NER lined Saxon green livery, and embellished with brass-capped chimneys. Almost immediately after the last had been completed, in June 1904 it was ordained that from henceforth lined black was to be the livery for all NER goods locomotives, which therefore included all the Darlington-built 0-8-0s.

The appearance of the first NER 0-8-0s reflected the rapid development of heavy industry in the North-east in the early 20th century, with huge quantities of raw materials, coal, coke, iron ore and limestone, being transported to steel works and sea ports, etc, and also across the Pennines over Stainmore to Penrith. The rather light-weight Thomas Bouch-designed Belah Viaduct en route could not have borne the concentrated weight of an 0-6-0 of sufficient power, which therefore dictated the spreading of some 60 tons of engine weight over four coupled axles.

The NER annually carried the highest mineral tonnage of any of its contemporaries. This increased by 26 per cent between 1900 and 1911, while the mineral train mileage that carried it decreased by no less than 38 per cent below its 1900 peak (see Table 19). Over the same period the revenue derived from this commodity rose by merely 15 per cent. The doubling in productivity between 1900 and 1911 must have been mainly due to the introduction of more powerful locomotives, i.e. the P2 and P3 large-boilered 0-6-0s as well as the Worsdell 0-8-0s. Disappointingly this could not have been attributed to the NER's attempts, like those of the CR, GWR and GCR in the same period, to adopt high-capacity bogie wagons, in place of the traditional 8- and 10-ton four-wheeled mineral wagons. These new wagons offered a greater payload per ton tare weight, hence the need for fewer wagons, thereby cutting building and repair costs. This also made for shorter trains, fewer axleboxes meant lower rolling resistance, and even clerical costs could be reduced. Regrettably the railway companies failed to persuade the colliery owners to invest in new screens in order to accommodate these new high-sided bogie vehicles, which in practice came to be used for transporting locomotive coal in bulk to locomotive depots, as this was a non-remunerative purely operational activity.

although the new 0-8-0 was simply the goods equivalent of the S2 mixed traffic engine, it actually proved to be considerably more successful and longer lasting. Although this was to have the same superheated boiler, it was provided with a deeper ashpan, and generally similar cylinders, although these were given improved exhaust ducts to the blastpipe. Ninety more T2 0-8-0s were subsequently built between 1917 and 1921, the last fifty by Armstrong Whitworth & Co. in Newcastle upon Tyne as its first locomotives following its turn over to producing these immediately after the Armistice in 1918. Whereas the 4-6-0s, built between 1911 and 1913, were all retired between 1937 and 1947, the invaluable 0-8-0s remained intact until 1960.

In 1902 only the Great Central Railway joined the 0-8-0 club, initially ordering just three engines for trial from Neilson, Reid & Co. These had many features in common with the 4-6-0s that were also being built alongside for the GCR. Not only that but both classes had clearly been inspired by the corresponding T and S classes on the NER respectively. Large numbers of 0-8-0s were subsequently built for the GCR by Kitson & Co. in Leeds, and latterly by the GCR at Gorton Works.

Despite the LNWR ultimately becoming by far the largest operator of 0-8-0s, Crewe did not actually turn out any new examples of these for five years between 1905 and 1909, although during this period many of the 280 existing Webb compound 0-8-0s were rebuilt in various

## Table 19 NER mineral train statistics, 1870–1911

| Year | Mineral tonnage (million) | Mineral train mileage (million) | Average tons/mile | Mineral revenue £ million | Revenue per mile p |
|------|---------------------------|----------------------------------|-------------------|----------------------------|---------------------|
| 1870 | 20.05 | 5.47 | 3.67 | 1.73 | 31.7 |
| 1880 | 29.11 | 6.84 | 4.27 | 2.49 | 36.4 |
| 1890 | 32.12 | 7.65 | 4.33 | 2.61 | 34.0 |
| 1900 | 40.55 | 8.01 | 5.06 | 3.06 | 38.2 |
| 1903 | 42.03 | 5.93 | 7.09 | 2.97 | 50.1 |
| 1906 | 46.58 | 5.78 | 8.06 | 3.29 | 56.8 |
| 1909 | 49.46 | 5.18 | 9.54 | 3.47 | 67.0 |
| 1911 | 51.06 | 4.94 | 10.34 | 3.51 | 71.1 |

Note: 1912 has been omitted on account of that year's miners' strike.

After a lull during 1912, Darlington Works then built thirty new 0-8-0s with 5ft 6in diameter superheated boilers (Class T2) in 1913. In 1966, five years before his death, Richard Inness recounted how, on descending seniority grounds, he had been deputed to prepare the detailed designs for these by the chief draughtsman (George Heppell).[1] Heppell's deputy (Ralph Robson) was already heavily engaged on designing the new Newport–Shildon electric locomotives, while the leading draughtsman (H. Spencer) was likewise preoccupied with redesigning the recent S2 4-6-0 for one to be fitted with Stumpf Uniflow cylinders (see Chapter 9). Essentially,

ways. This resulted in the appearance of the inside-cylinder, larger-boilered Class G in 1906 that set the basic pattern for all the rest still to come, and of which sixty new engines would later be built in 1910. These new engines were preceded by 170 19in goods 4-6-0s, derived directly from the Whale Experiment express passenger engines, and built between December 1906 and October 1909. No fewer than seventeen of these small-wheeled 4-6-0s were turned out during the single month of July 1908 alone. Rather curiously, no sooner had this class made its debut than there was also a proposal in April 1907 for an inside-cylinder 2-8-0, likewise with 5ft 2½in diameter coupled

[1] *Journal of the Stephenson Locomotive Society*, October 1975, p. 339.

A London & North Western Railway 19in goods 4-6-0 (c. 1907). Although very similar to the Experiment class express passenger engines, their respective boilers were not interchangeable. *Former Ian Allan Archive*

wheels. This would therefore have been something of a hybrid between the 19in 4-6-0 and the Class G 0-8-0, but in the event it was not proceeded with. The superheated Class G1 duly appeared in 1912, and once again the 'magic' total of 170 of these was built at Crewe up to 1918. The ultimate G2 class, of which sixty were finally built during 1921–22, differed from the G1 only in having direct Joy motion, similar to that being fitted to the latest 'Prince of Wales' 4-6-0s that were simultaneously being built in Glasgow under contract by William Beardmore & Co. The ultimate LNWR locomotive design consisted of a massive 0-8-4 tank version of the G2, of which thirty were built at Crewe 'posthumously' during 1923, particularly for service in South Wales.

Finally, as regards 0-8-0s that were actually built, in 1907 the Hull & Barnsley Railway took delivery of fifteen engines from the Yorkshire Engine Co. to handle its rapidly increasing export coal traffic from new pits in South Yorkshire to Alexandra Dock in Hull. Designated Class A, these were the largest locomotives to be designed by a member of the Stirling family, and the only ones to incorporate the Belpaire firebox, while still remaining true to the family's long-established domeless tradition. The original 0-8-0 boiler working pressure of 200lb literally frightened the engine crews, in view of the local boiler explosion at Knottingley on the LYR back in 1901, and another very recently involving an 0-6-2T on

the HBR itself. They successfully negotiated an increase in pay for working on these big engines, although the pressure was later reduced to 175lb. No further 0-8-0s were ordered by the HBR, which thereafter reverted to modestly proportioned 0-6-0s, the final five of which, delivered in early 1915, uniquely for a Stirling family design, were superheated.

There was a distinct hiatus in 0-8-0 construction during 1908–09, when there was a downturn in the national economy, when only thirty-three such engines newly appeared, and no GWR 2-8-0s were built. On the other hand, during the same period just over two hundred new 0-6-0s entered traffic, which included sixty P3 engines on the North Eastern Railway. However, during 1910–14 twice as many eight-coupled goods engines, now increasingly including 2-8-0s, were built, in each year consistently outnumbering new 0-6-0s. During the period 1908–14 there was a remarkable contrast between the LNWR and the GWR as regards heavy goods locomotive construction. Whereas Crewe built 115 19in 4-6-0s and 160 Class G/G1 0-8-0s, Swindon turned out a mere twenty 2-8-0s. Furthermore, early on during that period on the GWR there had been a perceived surplus of Dean '2301' 0-6-0s, when twenty of these, all dating from 1896, were somewhat improbably converted into (inside-cylinder) 2-6-2Ts between 1907 and 1910, in order to meet a shortage of suburban tank engines at that time.

## 2-8-0 locomotives

The most significant eight-coupled locomotive to emerge in Britain during 1901–14 was an initially solitary 2-8-0, numbered 97, completed by the GWR at Swindon Works in July 1903. Only the second new Churchward locomotive to carry a (half-cone) taper boiler this was a heavy goods equivalent of the Saint class express passenger 4-6-0, of which No. 98 had been completed just four months earlier. Although the first British 2-8-0, this wheel arrangement was extremely numerous in the USA, where it was known as the 'Consolidation' type, no fewer than 1,845 being built for the Pennsylvania Railroad alone between 1901 and 1913.

The production engines differed from No. 97 (later renumbered 2800) principally in having noticeably higher-pitched taper boilers that were now pressed to 225lb. On 25 February 1906 No. 2808 hauled a goods train aggregating 2,012 tons, including the Dean dynamometer car, between Swindon and Acton, 70 miles, at an average speed of just under 20mph.[2] With the progressive embellishment of superheaters (first fitted to the class in 1909), later full cone boilers, and characteristic Swindon top feed in succeeding batches, only fifty-six

engines were built up to 1913. The year 1914 saw a sharp decline in new eight-coupled locomotive building in general. Surprisingly, the GWR built no further 2-8-0s until 1918–19, after Nos 2800–55 had been heavily engaged throughout the First World War in working the famed but very rarely photographed Jellicoe Specials as far north as Warrington. These trains conveyed vast quantities of Best Welsh steam coal from the South Wales valleys through to Grangemouth, destined for the Grand Fleet based at Scapa Flow in the Orkneys. The Royal Navy's coal consumption at the peak of its operations in the North Sea amounted to no less than 100,000 tons per week. After a gap of almost twenty years a further eighty-three 2-8-0s, updated with outside steam pipes and Collett side window cabs, were later built at Swindon during 1938–42.

The former GWR 2-8-0s later proved to be the most economical in coal consumption when competing against other more recent 2-8-0 heavy goods designs, during the course of the British Railways locomotive exchanges in 1948. By the early 1950s the oldest members of the class had passed the forty-five-year mark, which was their theoretical economic life. British Railways planned to begin to replace these with the first batch of its new Standard

GCR Class 8K 2-8-0 No. 1196, one of the eighty-nine locomotives of this type delivered during 1912, this one being of Kitson manufacture, and seen at Mexborough. *A. W. Croughton/Rail Archive Stephenson*

Class 9F 2-10-0, which was scheduled to appear in 1953. The now Western *Region*, which entertained a rooted dislike of the new BR Standard locomotives in principle, made a reasoned case to be allowed instead to build some more of its home-grown 2-8-0s, claiming that these would be cheaper to operate. This proposal was firmly vetoed by higher authority, but as a result a BR Standard 8F 2-8-0 was proposed specifically for service on the Western Region in lieu. As outlined in simple diagram form in November 1953, this was either to be derived from the BR Standard Class 5 4-6-0, or amount to a truncated eight-coupled version of the BR 9F 2-10-0.[3] Five 2-8-0s, yet to be designed in detail, were initially authorised on the BR 1956 locomotive building programme, but were quickly replaced by a like number of 2-10-0s after all. Old GWR No. 97, as No, 2800, would soldier on and notch up *fifty-five* years' service, when in 1958 rather appropriately it became the first Swindon-built 2-8-0 to be withdrawn.

Further north only seven months elapsed between the completion of the last Robinson non-superheated 0-8-0 and the first superheated 2-8-0 at Gorton Works in 1911 (GCR Class 8K). Although a larger, superheated 2-8-0 indeed seemed to represent a logical progression, the Gorton drawing registers indicate that during July 1910, not only were larger 2-8-0s having either three cylinders, or 22½in diameter *inside* cylinders, but also a 2-8-2 and 2-10-0 were previously proposed.[4] Details of all of these schemes, prepared by one H. Lee, regrettably are now lost. The soon to be very familiar outside-cylinder 2-8-0 was drafted almost as an afterthought by an unidentified draughtsman nearly two weeks after the others on 2 August. This was approved almost immediately when twenty were ordered from Gorton Works only six days later. The rationale had been the anticipation of a massive increase in export coal traffic from South Yorkshire with the completion of the Great Central's massive new docks at Immingham on the North Sea coast. This facility was opened by King George V on 22 July 1912, almost exactly six years to the day after the ceremonial cutting of the first sod in July 1906. In addition to the first twenty 2-8-0s from Gorton Works, orders were also very soon placed with NBL in Glasgow for fifty, and with Kitson & Co. in Leeds for another twenty.

The Great Central 2-8-0 had a highly complicated secondary history. This was connected with the fact that the Great Central Railway's general manager, Sir Sam Fay, who had been knighted by the King at the opening of Immingham Dock, was seconded to the War Office during the coming conflict. At its behest the wartime Ministry of Munitions ordered a total of more than five hundred 2-8-0s of the Robinson design from four commercial locomotive builders, for service with the Railway Operating Division of the British Army in France. Somewhat unusually, twenty-five were ordered by the government from Gorton Works, although owing to the cessation of hostilities only six were completed in early 1919, of which three were retained by the GCR. Nevertheless, 2-8-0 construction elsewhere continued into 1920 despite some being cancelled, and by which time large numbers were dumped as surplus to requirements, with several being sold overseas to China and Australia. Many others remained, however, and were later purchased by the LMSR (75), GWR (90) and LNER (280). Those acquired by the LMS, however, were either scrapped or were otherwise disposed of by 1933, whereas many of those purchased by the LNER went on to be extensively rebuilt in a variety of ways, some with new LNER standard 100A boilers until as late as 1958, when the last examples acquired by the GWR were finally withdrawn.[5]

Between 1918 and 1921 the GCR built a further nineteen 2-8-0s fitted with 5ft 6in diameter boilers (Class 8M). Although very handsome engines indeed – the final five were even given side-window cabs – the extra 6in on the boiler apparently conferred no perceptible advantage, and two engines were converted to the standard 5ft variety as early as 1922. There were to have been twenty of these 2-8-0s, but one of the initial 1918 batch was instead turned out as a 5ft 8in mixed traffic 4-6-0 (Class 8N) as a possible future standard type. (See Chapter 10)

Both the Great Western and Great Central 2-8-0s, although having outside cylinders, had inside Stephenson valve gear. On the Great Northern Railway in 1913 Nigel Gresley made a bold move by building some 2-8-0s having external Walschaerts valve gear, which had already appeared on his first 2-6-0s built the previous year (see Chapter 10). The Gresley 2-8-0s looked rather clumsy affairs with their massive 21in by 28in cylinders and short-travel piston valves, although they did represent quite an advance over the earlier Ivatt 0-8-0s. A distinctly more refined *three*-cylinder version of the 2-8-0 was later developed at Doncaster from 1918 (Class O2), which would continue in intermittent production over the next twenty-five years.

In 1914 the Midland Railway at Derby built six 2-8-0s with 4ft 7½in diameter coupled wheels and, like the recent GNR engines, having massive 21in by 28in cylinders and outside Walschaerts valve gear, for the Somerset & Dorset Joint Railway. These were the sole outcome from a sequence of eight-coupled goods engines that had been drafted at Derby over several years both for the Midland Railway itself and the SDJR.[6]

[2] Carling, D. R., Two-thousand tons with one locomotive in 1906, *The Railway Magazine*, August 1969, pp. 439–40.
[3] Atkins, P. *The British Railways Standard 9F 2-10-0*, The Irwell Press, 1993, both proposed BR 2-8-0 diagrams are illustrated on p. 18.
[4] Atkins, P., Counsels of Perfection? – a new look at Robinson Great Central Railway locomotives, *Back Track*, August 2001, pp. 445–52.
[5] *Locomotives of the LNER*, Part 6B, Tender engines – Classes O1 to P2, The Railway Correspondence & Travel Society, pp. 33–75.
[6] Atkins, P., Midland eight-coupled, *Midland Record* No. 13, 2000, pp. 38–43.

*Above:* Great Northern Railway Class O1 2-8-0 No. 459, built at Doncaster Works in January 1914. *Former Ian Allan Archive*

*Below:* The first Somerset & Dorset Joint Railway 2-8-0 No. 80, newly completed by the Midland Railway at Derby Works in early 1914. For a numerically small class taken over by the LMSR, the design enjoyed the surprisingly long operational span of precisely fifty years, which ended at Bath in 1964. *Former Ian Allan Archive*

These almost exactly contemporary 2-8-0s were the first British heavy goods locomotives to feature (short travel) outside Walschaerts valve gear.

## 0-8-0s and 2-8-0s that might have been

The Midland Railway had begun to consider building some 0-8-0s for its own use as early as 1902, and drawings were later completed for an outside-cylinder locomotive with Belpaire firebox, which would otherwise have fairly closely resembled the North Eastern Class T engines, due to John W. Smith now being the chief draughtsman at Derby. Ten such engines were authorised in late 1903, shortly before the retirement of Samuel Johnson, only for these to be cancelled almost immediately. Johnson's successor, Richard Deeley, himself later proposed something very similar, but without success, and later two further 0-8-0 schemes for the steeply graded SDJR, which would seemingly have been fitted with an external version of his own valve gear. A superheated development with Walschaerts valve gear of one of these, provided with a leading pony truck, and designed by James Clayton, was built for the SDJR in 1914. In place of the usual 0-6-0s working in pairs, these 2-8-0s were later tried by the Midland on Toton–Brent coal trains, on which No. 85 was photographed in March 1918, followed by No. 83 in June 1919.[7]

The civil engineer, therefore, did not appear to entertain a rooted objection to eight-coupled locomotives in principle working on the Midland Railway. Indeed, in 1919 Derby mooted a larger-boilered version of the 2-8-0 for its own use, anticipating the five further 2-8-0s that were later built for the SDJR in 1925. In mid-1920 two much larger 2-8-0s still were proposed that would have utilised the new Lickey 0-10-0 boiler. As against the 66 tons of the SDJR engines, their estimated weights were 70½ tons (4ft 8½in coupled wheels) and 72 tons (with 5ft 3in). The latter diagram was tersely endorsed 'Too heavy', and may not even have been submitted to the civil engineer for his approval. The highly conservative Derby Class 4 0-6-0, dating from 1911, with its small axleboxes, would continue to be built under LMSR auspices for a further thirty years, despite periodic proposals to produce a more up-to-date replacement. The first was for an outside-cylinder 2-6-0 as early as 1920, which would have resembled a 'three-quarter' Somerset & Dorset 2-8-0. However, this objective was not finally achieved until 1947 by H. G. Ivatt with his austere Class 4 2-6-0 for the LMSR. This, it has been cogently argued, after all that time was not a significant improvement.[8]

In Scotland, the North British Railway prepared drawings, around 1908, for an 0-8-0 to handle its heavy Fife–Aberdeen coal traffic. Having outside cylinders, this would have borne a similar relationship to the Great Central 0-8-0s, as did the NBR 4-4-2s to the likewise somewhat smaller Gorton 'Atlantics'. The NBR 0-8-0 was not immediately proceeded with owing to the prevailing adverse economy; it was possibly nearly revived around 1912, yet still did not eventuate. In 1916 the North British tested an NER T2 north of Dundee, and later in 1921, on different occasions, it 'hosted' a GWR 2-8-0, No. 2804 (in appalling winter weather conditions), and later on an NER three-cylinder T3, for evaluation on the testing gradients at Glenfarg near Perth. As a result, the NBR tentatively proposed a rather heavy three-cylinder 2-8-0, but by then it was too late in the day, following the almost simultaneous promulgation in Parliament of the legislation for the forthcoming railway amalgamations.

Two other more surprising railways also proposed 0-8-0s very early in the 20th century. In October 1901 James Clayton, in the Ashford drawing office of the SECR, got out two alternative schemes for an inside-cylinder 0-8-0 for Harry Wainwright. The definitive history of SECR locomotives states that the construction of ten 0-8-0s was authorised in March 1902, but that ultimately an equal number Wainwright Class C 0-6-0s was built instead.[9] The 0-8-0 would have carried the Class D boiler, but in contrast would have looked distinctly ungainly, having none of the elegance of that 4-4-0, to which Clayton had also contributed, nor the easy charm of its sister 0-6-0.

In late 1902, more detailed proposals were prepared in Inverness for Peter Drummond by the Highland Railway's chief draughtsman, Robert Collie, for a large inside-cylinder 0-8-0. The boiler would have been of generous proportions with an outside diameter of 5ft 3in, and a grate area of 27 sq ft. No indications were given of any estimated weights.[10] When serving both the HR and later the GSWR, Peter usually slavishly followed his elder brother Dugald's locomotive designs for the London & South Western Railway, but there had been no English precedent in this instance.

Just ten years later, Dugald Drummond would die suddenly in early November 1912 as a consequence of an untreated accidental scalding. Only about three months earlier, when he had instigated design work at Eastleigh on yet another four-cylinder 4-6-0 for the LSWR, a four-cylinder 0-8-0 heavy goods, Class H15, with 5ft 1in coupled wheels was also designed in parallel. The two classes were to have several major features, including boilers and cylinders in common. Of these, the boiler was to revert to the large 5ft 9in diameter pattern of the earliest LSWR 4-6-0s, and serving the 16½in cylinders would have been extremely generous 10in diameter piston valves as on the previous Class T14. LSWR locomotive classifications were actually alphanumeric works order numbers, instituted by William Adams, indicating that these engines, probably five of each, had already been

[7] *Ibid*.

[8] Tester, A., *A Defence of the Midland/LMS Class 4 0-6-0*, Crimson Lake, 2012.

[9] Bradley, D. L., *The Locomotive History of the South Eastern & Chatham Railway*, The Railway Correspondence Society, 2nd edition 1980, pp. 133/9.

[10] Cormack, J. R. H., & Stevenson, J. L., *Highland Railway Locomotives*, Book 2, The Railway Correspondence & Travel Society, 1990, pp. 154–5.

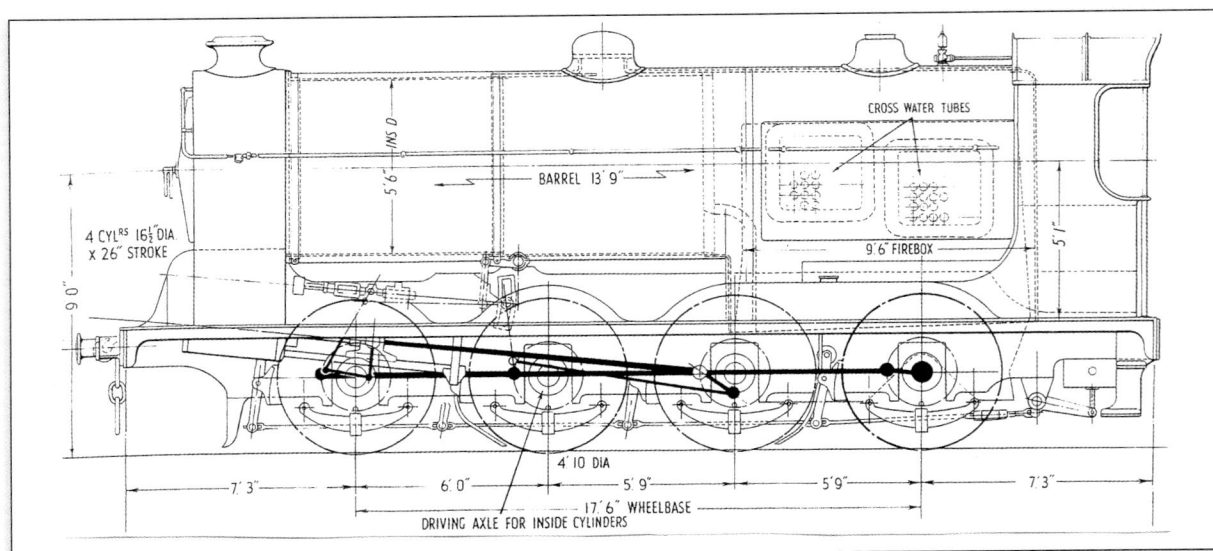

Diagram of proposed D. Drummond LSWR four-cylinder 0-8-0 (1912).

authorised before Drummond's unexpected demise. In fact the boiler barrels had already been fabricated. The final 'joint' entry in the Eastleigh drawing register relating to Classes H15/K15 was dated 23 October 1912, which must have been roughly coincident with the beginning of Drummond's sudden unscheduled absence, from which in the event he would not return. However, while this marked the end of the 0-8-0, this was by no means the end of the matter as far as the 4-6-0 was concerned, which later appeared in a very different form, as will be described in Chapter 10.

Given that no more 0-6-0 goods engines were built for the LSWR after 1897, the preponderance of 4-4-0s of various classes, and the nature of the goods traffic on the LSWR in general, the intriguing question is what was the actual justification for Drummond's extremely large 0-8-0? Interestingly, his successor, Robert Urie, for his part allegedly proposed a two-cylinder 4-8-0, also with 5ft 1in coupled wheels.[11, 12] The most plausible answer is that both were intended for operations associated with the new marshalling yard at Feltham, in south-west London, which had been planned immediately pre-1914 and completed very soon after the war had ended. As it was, two closely related tank locomotive classes were built at Eastleigh in 1921 specifically for operating there, i.e. five powerful 4-6-2Ts for local transfer freight workings, together with four massive 4-8-0Ts for heavy shunting duties.

## Proposed ten-coupled heavy mineral locomotives, 1914

The outbreak of the First World War in August 1914 frustrated two very interesting *ten*-coupled locomotive projects, on the Lancashire & Yorkshire and Great Central railways, both of which had respectively been carrying large quantities of export coal eastward from Lancashire to Goole, and from the South Yorkshire and Nottinghamshire & Derbyshire coalfields to Immingham.

In late 1913, at the request of the LYR's general manager and former CME, Sir John Aspinall, the locomotive department under George Hughes began investigations into how the company's heavy coal trains might be worked more economically, especially to Goole. In 1913 this particular traffic amounted to 1.4 million tons, compared with 1.6 million tons in 1910. (Interestingly *canal*-borne coal to Goole amounted to 1.6 million tons in 1913, and 1.3 million tons in 1910.) Six outline schemes were thereafter put forward:

Superheated four-cylinder compound 2-8-0
(17½-ton axle load, two schemes)
Superheated four-cylinder simple 2-8-2
(19½-ton axle load, three schemes)
Superheated four-cylinder simple 2-10-0
(16-ton axle load, one scheme)

The two compound 2-8-0 schemes would have somewhat resembled a French PLM compound 2-8-0 that was

[11] Bailey, P., The Urie 4-8-0, *The Southern Way*, No. 3, April 2007, pp.23–5. This engine would have been very closely based on the Urie H15 class 4-6-0.

[12] In April 1929 Eastleigh drawing office produced a chart of Southern Railway standard locomotives. This included a proposed 4-8-0 mineral engine with 5ft 1in coupled wheels, which was directly derived from the recent four-cylinder 'Lord Nelson' express passenger 4-6-0. What had prompted this proposal also remains unknown. In *Locomotive Adventure* (1962), p. 178, Harold Holcroft remarked 'when the matter was raised with the Traffic Department they replied that they had no use for a larger freight engine; existing types could cope with the heaviest trains they found desirable or practical to run, in view of track occupation and the absence of running loops.'

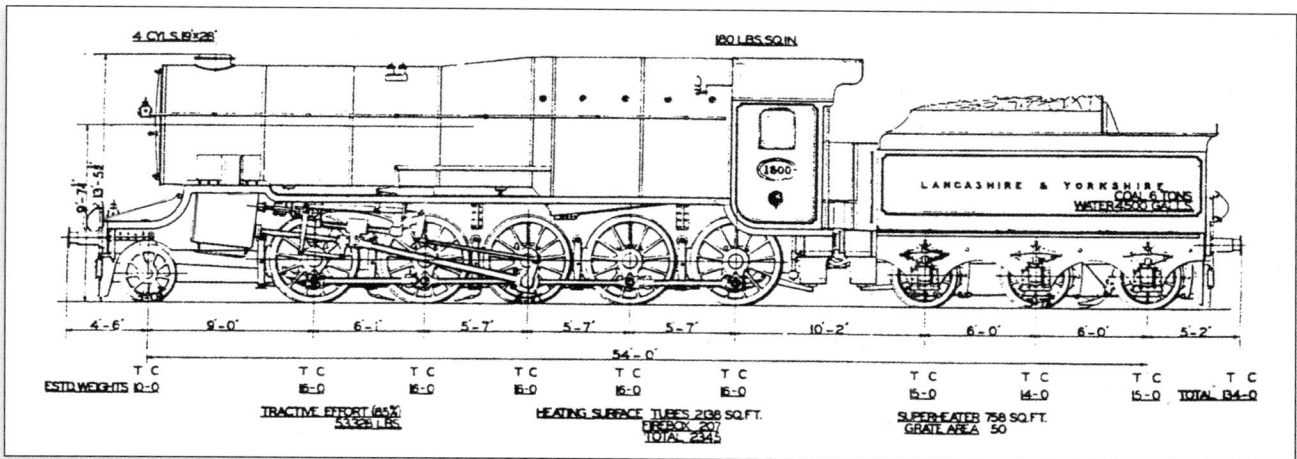

*Above:* Drawing of proposed LYR four-cylinder 2-10-0, 1913–14.

*Below:* Diagram of proposed Great Central Railway two-cylinder 2-10-2, May 1914.

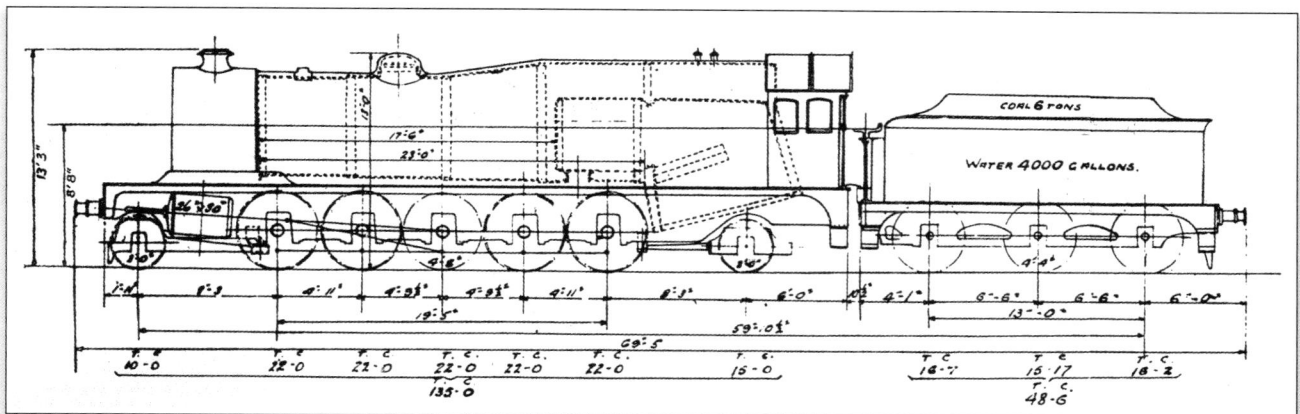

exhibited at Ghent in 1913, and the LYR proposals called for inside low pressure cylinders of no less than 25½in diameter, which would have been extremely difficult to accommodate satisfactorily between the frames with their traditional inside spacing of 4ft 1½in. In the event, the 2-10-0 option was selected and worked up in some detail, the dimensioned weight diagram being dated June 1914. Apparently it would even have embraced the enlightened intention of providing exceptionally long travel (7½in) piston valves. The shallow firebox affording a hitherto unprecedented grate area of 50sq ft would have made for extremely hard work for the fireman, whose stamina and skill would have been key to the desired economy. Nearly ten years later the LYR 2-10-0 scheme was evidently briefly reappraised by the newly formed LMSR, possibly for Toton–Brent, but if nothing else the unusually generous maximum height of 13ft 5½in to the top of its already practically non-existent chimney could well have presented major clearance problems.

The overall concept had clearly been inspired by the slightly larger and heavier Belgian four-cylinder 2-10-0s designed by J. B. Flamme, and introduced in 1910, which

a delegation of LYR officers, including George Hughes, would have inspected during a visit to Belgium soon afterwards.

Particularly boiler-wise, the Flamme 2-10-0 would also appear to have inspired a Great Central proposal for a two-cylinder 2-10-2, the date of the initial diagram being 20 May 1914. This monster would have had two 26in by 30in cylinders, and an adhesive weight of 110 tons, although at the time neither of these extreme dimensions was perceived as posing a particular problem on the Wath–Immingham route, other than requiring the opening out of a very short tunnel at Conisborough.[13] As with the LYR 2-10-0, the GCR 2-10-2 would likewise have been paired with a disproportionately small existing company standard six-wheeled tender.

It is interesting to note that a few years later, in 1922, there was a serious proposal by Nigel Gresley on the Great Northern Railway to build three-cylinder 2-10-2s, derived from his new 'Pacifics', to work the heavy Peterborough–London coal trains. This later appeared *in spirit* in 1925, as the two LNER Class P1 2-8-2s with eight 5ft 2in in place of ten 4ft 8in diameter coupled wheels. Like the

[13] Dow, G., *Great Central*, Vol. 3, Ian Allan, 1963, pp. 323–5.

Caledonian 0-8-0s, these could haul longer and heavier trains than the existing track infrastructure could accommodate. In the event, the special 'Lickey' 0-10-0 apart, a purely domestic British ten-coupled goods tender locomotive did not finally make its appearance until the eleventh hour, in 1954, in the form of the excellent British Railways Standard Class 9F 2-10-0. In purely practical terms the latter could be said to have been the first quantum leap in British conventional goods locomotive design for fifty years, since the emergence of GWR 2-8-0 No. 97 back in 1903.

## Survival of the fittest

In marked contrast to their 4-6-0 passenger contemporaries, many Edwardian eight-coupled heavy goods engines lasted well into the 1960s. The relative longevity, or otherwise, of the superheated 0-8-0s and 2-8-0s that were all built during the year 1913, when examples were built by six different railways, makes for interesting comparisons:

LNWR G1 0-8-0 (24), unrebuilt,
    withdrawn 1947–51
as rebuilt to G2A 1957–62
LYR 0-8-0 (31) 1930–50
GNR O1 2-8-0 (3) 1951–52

GCR 8K 2-8-0 (13), unrebuilt 1962
    as variously rebuilt 1962–65
NER T2 0-8-0 (30) 1963–67
GWR 28XX 2-8-0 (6) 1960–63

In terms of what proportion still remained twenty-five years after 1922, the locomotives of the former LYR *in general* fared distinctly better at 36 per cent, than did those of the LNWR at only 23 per cent. However, the reverse was true in the case of their respective 0-8-0s. For example, thirty-four of the last forty-three 0-8-0s that were built at Horwich during 1919–20 were withdrawn between 1931 and 1939. By contrast, their Crewe close contemporaries, the sixty G2s built in 1921–22, remained intact until 1959. Withdrawal of the Gresley two-cylinder O1 2-8-0s, of which twenty in total were built between 1913 and 1919, surprisingly began as early as mid-1947, actually starting with three of the later 1919 NBL-built engines. On the other hand, new boilers were still being built for the former NER T2 0-8-0s at Darlington Works remarkably until as late as 1960. This was possibly by way of an alternative to building some additional new Class 9F 2-10-0s, for which allegedly a post-1955 BR Modernisation Plan request was made by BR North Eastern Region, but which had been rejected by higher authority. These were required for short-haul coal train working in County Durham, when there was no economic diesel-electric locomotive alternative available.

Great Northern Railway (USA) Class G5 4-8-0 heavy goods engine built in 1897 by Brooks, who then billed it as being 'The Largest Locomotive in the World'. This incorporated the fundamental characteristics, i.e. tapered Belpaire boiler and distinctive cylinders, attributed to the Welsh-born John Player, which undeniably inspired George Churchward on the GWR which were perpetuated in new construction at Swindon for almost fifty years. Making an interesting comparison with the GWR 2-8-0s illustrated on p. 81, the 4-8-0's 21in diameter cylinders were served by exceptionally large 16in diameter piston valves, and they also had the then unprecedented piston stroke of no less than 34in. It is also interesting to note that its grate area of 34 sq ft was only equalled in a narrow firebox in British locomotive practice on the GWR King class 4-6-0s, that were introduced thirty years later. However, after 1900 in the USA, narrow fireboxes only rarely featured in new locomotive designs, being necessarily superseded by much larger wide types.

# 9
# Cutting the Coal Bill

Coal was the lifeblood of the railways, many of which also derived considerable revenue from transporting it. In 1913 the movement of coal accounted for £22.7 million, or 19 per cent of the railways' combined traffic gross revenue. Also during 1913, when the combined annual output of British collieries reached its all-time peak, this came at a huge human cost. On 14 October multiple pit explosions at Senghenydd near Caerphilly in South Wales resulted in the loss of 439 lives. Although the railways themselves collectively consumed 13.6 million tons of coal per annum, this nevertheless ranked the railways very far down the domestic pecking order. This was headed by a huge manufacturing industry at 75 million tons, followed by the then very heavy demands of the domestic hearth (40 million tons), and it was even well behind that of the coal industry itself (18 million tons) in fuelling its winding engines to raise the commodity to the surface. Although the value of the pound sterling had hitherto remained virtually unchanged for decades, the cost of coal per ton could vary quite significantly even from year to year according to demand and thereby profoundly affect the profitability of the railways. When considered overall, fuel accounted for about half of the running costs of the national locomotive fleet, and in 1913 the expenditure on coal was roughly equivalent to 10 per cent of the total gross traffic revenue that was generated from burning it in locomotive fireboxes. Therefore, any means by which coal consumption might be reduced in some way was keenly appraised by various railway companies, large and small. The LNWR attracted the largest fuel bill, which by 1913 now exceeded £1 million per annum.

## Compounding

The loss of its coalfields when Alsace–Lorraine was ceded to Germany after the Franco-Prussian War in 1871 prompted France to adopt *compound* locomotives (which expanded steam *twice* prior to exhaust) on a large scale, in order to reduce the quantity of coal now having to be imported from elsewhere, particularly Britain. Even after that region was restored to France in 1919 in accordance with the Treaty of Versailles, four-cylinder compound locomotives continued to be developed with a very high degree of sophistication on account of the still continuing necessity to import large quantities of locomotive coal. A final advanced range of proposed high-powered *three-cylinder* compounds, largely designed during the Second World War under André Chapelon, regrettably never progressed beyond the drawing board, owing to the adoption there instead of an alternative post-war policy of extensive main line railway electrification.

In Britain in the late 19th century, however, only two railways, the London & North Western under Francis Webb, and the North Eastern under Thomas Worsdell, built any significant numbers of compounds in a bid to reduce operating costs, albeit by means of more complicated locomotives to maintain. There was a direct link here, Worsdell had been works manager at Crewe for several years when the first Webb three-cylinder 2-2-2-0 compounds were built there in 1880. In 1882 he had moved on to become locomotive superintendent of the Great Eastern Railway, on which he built twenty two-cylinder compound 2-4-0s during 1882–83, followed by eleven very similar 4-4-0s during 1884–85. After only three years, in May 1885, Worsdell moved on again, this time to the North Eastern Railway where he built 250 likewise inherently lop-sided two-cylinder compound locomotives of various types, passenger and goods, which ranged from 4-2-2 tender to 0-6-2 tank, before he retired on health grounds in 1890, although he was still retained as a consultant by the NER.

Meanwhile, back at Crewe, in the early 1890s Francis Webb began to build highly idiosyncratic three-cylinder compound 2-2-2-2 passenger and Class A 0-8-0 goods engines, but after 1896 he began to favour *four-* rather than three-cylinder compounds, the last example of which was an 0-8-0 built in 1900. A total of 240 new four-cylinder compound 4-4-0s, 4-6-0s and Class B 0-8-0s were

London & North Western Railway Class B four-cylinder compound 0-8-0, of which 170 were built at Crewe Works between 1901 and 1904. *Former Ian Allan Archive*

then built at Crewe between 1901 and 1905 alone, a number of which were still on order before Webb's retirement in mid-1903. In all of these the boiler diameter was increased from 4ft 2in to 4ft 6in, but the distinctly limited standard grate area of 20½ sq ft still remained. This final and quite considerable tranche doubled Webb's overall tally of compounds to almost five hundred, which never actually existed simultaneously as such. There had not been any real method by which their possible coal savings could directly be assessed, although the first Webb 4-4-0, No. 1501 *Iron Duke*, had originally been built as a four-cylinder simple engine in April 1897. Only thirteen months later it was renamed *Jubilee* coincident with its very early conversion to a compound, in order to conform with its companion No. 1502 *Black Prince*, which had been built three months later in July 1897. Rather strangely, no direct comparison appears to have been made between the two engines in their respective as built forms, although Webb later reported that up to the end of February 1899 they had collectively run 190,324 miles on a distinctly unrevealing *overall* average coal consumption, as both simple and compound, of 40.3lb per mile.

The official Crewe portrait of the first Webb four-cylinder compound 4-6-0 (1903) which was directly adapted from the 0-8-0 above. The final engine, completed in 1905, was also the first to be withdrawn only eight years later, in 1913. *Former Ian Allan Archive*

LNWR Alfred the Great class 4-4-0 No. 1942 *King Edward VII*, believed to be at Rugby, c. 1904. *Rail Archive Stephenson*

The forty LNWR engines of the Alfred the Great class, built at Crewe between 1901 and 1903, were slightly larger versions of the Jubilee class. Following the rapid 'flood' of the new Whale passenger 4-4-0s and 4-6-0s from 1904–05, their primacy was very brief, but two of the class did accomplish a surprisingly little known feat when new. On 19 June 1903, two Alfreds built just four months earlier, Nos 1965 *Charles. H. Mason*, and 1966 *Commonwealth*, worked a special Glasgow-bound train *non-stop* for 299 miles from London as far as Carlisle. The train departed Euston at 3.45pm and arrived at Carlisle two minutes ahead of schedule at 9.43pm. Each engine had carried a relief fireman, but little else is recorded regarding any other special arrangements that might have been made in order to achieve this remarkable feat. Such was repeated, but in the up direction, four months later on 8 October, when two Alfreds also worked the Royal Train non-stop from Carlisle to Euston.

Even while the final Alfreds were still under construction and Webb was still in charge, an existing engine, No. 1952 *Benbow*, was provided with independent sets of Joy valve gear for its high- and low-pressure cylinders. On test in September 1903 when the original arrangement was simulated this produced a maximum of 835 IHP, but when the respective valve gears were operated independently this was increased to 949 IHP. Commencing in 1913, the majority of the Alfreds were further rebuilt into Renown class two-cylinder simples, although inexplicably, while these conversions were still ongoing, one engine, No. 1974 *Howe*, uniquely was superheated in 1921 while still remaining a compound.

Webb retired on health grounds in mid-1903 and was succeeded by George Whale, who almost immediately began to scrap the controversial 2-2-2-0s, while the following year he commenced converting the ponderous three-cylinder 0-8-0s to two-cylinder simple expansion. This operation was completed in 1912, when the last remaining examples of the comparatively recently built (1898) three-cylinder 2-2-2-2s were scrapped. Remarkably, the first Webb three-cylinder compound 0-8-0, which had been built in September 1893, after being rebuilt on three successive occasions would remain in active service on paper at least until December 1962, by which time it is doubtful whether very much of the original engine could still have remained. Later, the four-cylinder compound 4-4-0s and 0-8-0s also began to be converted to simples by removing the outside high-pressure cylinders and reducing their boiler pressure from 200 to 175lb. Two of these 0-8-0s would have the distinction of being among the last of more than five hundred LNWR-built 0-8-0s to remain in service, lasting until December 1964. A few four-cylinder engines in the event were never rebuilt and in this form the last remaining Webb compound 4-4-0s and 0-8-0s were finally retired by the LMSR in 1928.

Very shortly after Whale's own retirement in mid-1909, *The Railway Magazine* for August 1909 featured an article on his locomotives by one J. N. Jackson (London & North Western Railway).[1] For some reason the journal coyly declined to identify his actual status, although John Jackson was in fact none other than currently the chief draughtsman at Crewe, which indeed he had been for twenty-one years. He would later retire in 1919, no fewer

---

[1] Jackson, J. N., The Whale Locomotives on the London and North-Western Railway, *The Railway Magazine*, August 1909, pp. 130–8.

than forty-four years after he had first entered the drawing office, and would therefore have been intimately involved in the design of all the Webb compounds. On the question of these he diplomatically had this to say:

'No man has done more to show what a compound locomotive can do, and also what it cannot do than Mr Webb … During the latter part of Mr Webb's regime at Crewe, the weights and speeds of passenger trains increased to such an extent that the Compounds were called upon to do more single-handed than they were able, with the result that it became necessary to put two engines to haul the main line express trains, a practice which certainly could not be defended on economic grounds.'

Mr Whale, on his succeeding Mr Webb, in 1903 reversed the policy of his predecessor, and being satisfied that the Compound system was not the best to deal with the conditions that had to be met in the heavy traffic on the London & North Western Railway, decided in favour of the simple system.

Significantly, John Jackson said nothing about fuel consumption, but remarked that the *repair* costs of the conversions to simple expansion (i.e. 0-8-0s) had been substantially reduced, not least due to the lowering of their boiler pressure. For their part, no alterations were made to the Webb 4-6-0s, and outright scrapping of these began with a 1905-built example as early as 1913. However, soon afterwards in early wartime a halt was called and there were tentative proposals to also rebuild these to simple expansion in a similar manner to the 4-4-0s and 0-8-0s. Thankfully, not least for appearances' sake, in the event they remained unaltered, and while their operation was undoubtedly prolonged by the war, the class finally became extinct in 1921.

There was a strong association between the London & North Western and Lancashire & Yorkshire railways, which for more than fifty years sought to merge, only to finally achieve this goal at the eleventh hour (and 59th minute!) in 1922. Locomotive-wise both extensively employed Joy valve gear from the 1880s until 1922, and in 1901 Henry Hoy rebuilt a ten-year-old Aspinall 4-4-0, No. 1112, as a four-cylinder compound that in addition also cosmetically closely resembled its Webb counterparts on the LNWR. Little is known of its performance before it was converted back to simple expansion again and superheated in 1908. In February 1906, Hoy's successor, George Hughes, rebuilt an only fifteen-month-old Aspinall 0-8-0, No. 1482, as a compound, with all four cylinders driving onto the second coupled axle. Despite concurrent developments on the LNWR, in 1907 he then built ten *new* compound 0-8-0s in which the outside high-pressure cylinders now drove onto the third coupled axle.

A detailed evaluation of all eleven compounds pitched against a like number of simple 0-8-0s was made between November 1908 and November 1909.[2] The compounds returned an average coal consumption of 88.4lb per *train* mile, as against 95.3lb per mile by the simples, a saving of 7.3 per cent although, as Hughes pointed out, there was no significant difference if expressed on a lb per *engine* mile basis, i.e. 59.4 v 60.7lb, which also encompassed all duties, e.g. shunting, light engine working, etc. This indicated that in view of the general rough and tumble of everyday service, the additional initial capital cost of building a compound, and then the ongoing extra maintenance costs associated with four as against only two cylinders, could not be justified.

No more compound 0-8-0s were built by the LYR and the next development came in July 1910 when further new inside-cylinder 0-8-0s began to appear now fitted

Lancashire & Yorkshire Railway four-cylinder compound 0-8-0 No. 974, built at Horwich Works in June 1907. Note the rigid frame eight-wheeled tender, whose use on the LYR was confined to its 0-8-0s, in view of the nature of their duties. *Former Ian Allan Archive*

[2] Hughes, G., Compounding and Superheating in Horwich Locomotives, *Proceedings of the Institution of Mechanical Engineers*, 1910, Vol. 78, pp. 399–451.

with larger diameter (5ft 9in) boilers. These were not as yet superheated, despite a production batch of twenty superheated 0-6-0s having already been built during 1909. *Superheated* large-boilered 0-8-0s then began to follow somewhat later in November 1912, in which their cylinder diameter was increased from 20 to 21½in, thereby rivalling the almost exactly contemporary new Great Central 4-6-0 No. 423 *Sir Sam Fay* on that particular score. Eric Mason recorded that it was originally intended that a batch of these 0-8-0s built during the winter of 1913–14 (LYR Nos. 1576–85) were initially intended to have been compounds.[3] This proposal could have been prompted by the recent very significant increase in the price of coal.

Long before this, the North Eastern Railway had already decided twenty years earlier that its *two*-cylinder compounds were not worth the candle, following the findings of a report dated November 1893 that had been commissioned by Thomas Worsdell's younger brother, Wilson, who had succeeded him as CME in 1890. While crediting the compounds with a slightly lower fuel consumption, the report stated that this was outweighed by their higher maintenance costs and the difficulties that drivers experienced restarting 0-6-0s, in particular when performing shunting operations. Conversion from compound to simple expansion began with the Class J 4-2-2s in 1895, and was completed on the last remaining 0-6-0 compound in 1910. However, at the insistence of the NER Board, against his own inclinations Wilson Worsdell completed a solitary two-cylinder compound 4-4-0 in May 1893, Class M 4-4-0 No. 1619. Following severe collision damage sustained five years later, this was then completely renewed as a *three*-cylinder compound, according to a system devised by the Gateshead chief draughtsman, Walter Mackersie Smith.[4] This consisted of one middle high-pressure cylinder exhausting into the two outside low-pressure cylinders, the complete reverse of the cumbersome Webb arrangement with its exceptionally large diameter (30in) middle cylinder. The Smith compound system was also designed to take live steam direct from the boiler at reduced pressure for admission to the low-pressure cylinders for starting purposes, or when working on a heavy gradient, as on Cockburnspath bank north of Berwick, which was referred to as 'reinforced compound working'.

No. 1619 was not developed any further on the NER, but W. M. Smith had a long-standing association with Samuel Johnson, who had been locomotive superintendent of the Midland Railway since 1874. Their connection began in the mid-1860s when Johnson had held a similar position on the Edinburgh & Glasgow Railway at Cowlairs, and Smith had been working for Neilson & Co. nearby in Glasgow. Smith's eldest son John, having served his time at Gateshead Works, then worked for a time in the drawing office there under his father before leaving in 1891 to work on piston valve development (of which Smith senior was an active pioneer) for Johnson at Derby. John Smith later rose through the Derby drawing office ranks to be appointed chief draughtsman in January 1901, and was undoubtedly closely involved in Johnson's decision to build some large three-cylinder compound 4-4-0s operating on Smith's father's system.[5] Five such were authorised in early 1900, perhaps significantly at a time when there was a temporary very sharp rise in the price of coal after the previous year's unprecedented industrial boom. The first two engines, MR

Midland Railway Johnson three-cylinder compound 4-4-0 No. 2634, built at Derby Works in 1903. Note the double bogie tender. *Former Ian Allan Archive*

[3] Mason, E., *The Lancashire & Yorkshire Railway in the Twentieth Century*, Ian Allan Ltd, 1954, p. 146.
[4] Smith, W.M., BP 14,721/1898 (relating to three-cylinder compound locomotives).
[5] Atkins, P., The Smith Connection, *Midland Record* No. 10, 1998, pp. 15–9.

The Deeley version of the Johnson compound 4-4-0, represented by MR No. 1030. Apart from technical changes, its more austere appearance is immediately apparent. *Former Ian Allan Archive*

Nos 2631/2, were completed in 1902, followed by three further engines, Nos. 2633–5, in 1903 in which the independent reversing gears for the high- and low-pressure cylinders (which could nevertheless be linked together) on the first two engines were replaced by a simple unified reverser. In other words, this was the very opposite of exactly contemporary developments on the LNWR four-cylinder compound 4-4-0s.

At a time when 4-4-2s and 4-6-0s were still extremely few and far between, these compounds ranked as the heaviest and most powerful British 4-4-0s so far built at 59 tons, and with a maximum axle load of 19½ tons. Johnson retired at the end of 1903 and his successor, Richard Deeley, built forty further extensively modified three-cylinder compound 4-4-0s between 1905 and 1909 in which the boiler working pressure was substantially increased from 195 to 220lb and the grate area from 26.0 to 28.4 sq ft. In these Deeley also eliminated the troublesome reducing valves of the Johnson engines, from which these were duly removed, and substituted a new starting arrangement involving a newly patented regulator valve.[6]

During December 1905 Gorton Works turned out a pair of Smith three-cylinder compound versions of the Great Central two cylinder 4-4-2 (GCR Class 8D), so designed as to be readily convertible to the standard type if desirable. These were followed by a further pair in December 1906 (Class 8E, although not immediately distinguishable from class 8D), and all four were subsequently given names, unlike the two-cylinder simple engines. Interestingly, between the appearances of these two small batches, John Smith departed Derby to become works manager at Gorton in August 1906. This post on the Great Central had formerly rather onerously been combined with that of chief draughtsman, which by convention elsewhere was considered to be the senior of the two positions. Smith's right-hand man at Derby, leading draughtsman James Anderson, had applied for the latter post on the GCR, but he was deterred from accepting it by what he considered to be the insufficient £450 salary on offer.

In 1908 John Robinson revealed that, compared with the two-cylinder engines, the four compound 4-4-2s were only slightly lighter on coal to the tune of 2–2½lb per mile, seemingly without actually stating what the respective figures were. Whatever, it would have been scarcely enough to justify their additional first cost and subsequent extra maintenance, yet these engines were never converted to standard, as was a two-cylinder 4-4-2, No. 1090, in 1922 several years after it had been

[6] The detailed development of the Derby three-cylinder compound 4-4-0s, under S. W. Johnson and R. M. Deeley, is expertly chronicled by Stephen Summerson in Volume 4 of *Midland Railway Locomotives*, The Irwell Press, 2005.

experimentally rebuilt as a three-cylinder *simple* in 1908. All four compounds were later routinely rebuilt with superheated boilers between 1911 and 1923, and would all remain in service until 1947, after achieving respectable working lives of forty years apiece.

It is unclear whether it had been simple fuel economy or performance that had been the incentive to build these four engines, which came towards the end of something of a four-year 'spate' of cylinder compound 4-4-2s, both actual and proposed, that were produced on the part of five English and Scottish railways:

Like the Pennsylvania Railroad at this time, George Churchward on the GWR was also minded to purchase an example in 1903 so that he could evaluate it and make a comparison with his own new locomotives. By 1905 he had demonstrated their superiority, and so it is far from clear why therefore, *two* further French compound 4-4-2s, this time of the larger Paris–Orleans design, were also purchased then.

Although the two NER compound 4-4-2s did not appear until mid-1906, design work upon them had begun at Gateshead in early 1903, i.e. well before the arrival of

## Table 20 British compound 4-4-2 locomotives, built 1903–07

| Rly | Date | Loco Nos | Builder | Notes |
|---|---|---|---|---|
| GWR | October 1903 | 102 | SACM* | French Nord design |
| GNR | March 1905 | 292 | Doncaster Works | H. A. Ivatt design |
| CR | April 1905 | – | (St Rollox Works) | Proposal only |
| GWR | June 1905 | 103/4 | SACM* | French P-O design |
| GNR | July 1905 | 1300 | Vulcan Foundry | Maker's design for GNR |
| GCR | December 1905<br>December 1906 | 258/9<br>364/5 | Gorton Works | J. G. Robinson design<br>(W. M. Smith system) |
| NER | April/May 1906 | 730/1 | Gateshead Works | W. M. Smith design |
| GNR | August 1907 | 1421 | Doncaster Works | H. A. Ivatt design |

* Société Alsacienne de Constructions Mécaniques, Belfort.

After 1900 it is more than likely that the incentive for these was to produce a potentially more powerful locomotive within prevailing constraints of maximum permitted weight, after the remarkable early performances, c. 1900–01, of the new de Glehn four-cylinder compound 4-4-2s on the Nord railway in France.

No. 102 *La France* on the GWR. John Smith had accompanied the Midland Railway 4-2-2 No. 2601 *Princess of Wales* when it was exhibited at the Paris Exhibition in 1900, as was NER 4-6-0 No. 2006 designed under his father. Also on display had been the first de Glehn compound 4-4-2 for the Nord, and John had taken the

NER Class 4CC 4-4-2 No. 730 after superheating in 1915, and the fitting of a plain chimney. Seen at York.
*Rail Archive Stephenson*

opportunity to ride on the distinctly minimal footplate of its sister engine. He would have undoubtedly reported back to his father at Gateshead, who took no further interest in three-cylinder compounds and moved on to *four*-cylinder compound locomotives, relating to which he quickly obtained two patents.[7] The forthcoming 4-4-2s, which most unusually would be officially attributed by the NER to W. M. Smith rather than Wilson Worsdell, incorporated both of these, which concerned the high- and low-pressure cylinders, and their associated piston valve and valve gear arrangements. On each side of the engine adjacent high-pressure and low-pressure cranks were set at 180 degrees to each other. The outside HP piston valves had inside admission, and the inside LP valves had outside admission so that each adjacent pair of valves were worked together in unison by one set of internal valve gear. This ingenious arrangement, however, precluded the luxury of independent cut-offs in the HP and LP cylinders, while the number of fixed cut-off combinations was limited to five by the use of steam-assisted lever, rather than screw reverse. The available HP/LP cut-offs (%) on No. 730 therefore were 41/54, 53/63, 59/66, 63/70 and 75/80 (full gear). The valve gear was Stephenson on No. 730, and Walschaerts on No. 731. Uniquely for NER locomotives, these engines sported Belpaire fireboxes, probably as a result of feedback from John Smith at Derby, where such had been adopted by the Midland Railway firstly on 4-4-0s in 1900. The 4-4-2s' unusual safety valves and their casing were an exact copy of those on the Johnson compound 4-4-0s.

Also during October 1906, No. 730 competed against other recent NER 4-4-0, 4-4-2 and 4-6-0 locomotives, and then against 'sister' No. 731 between October 1907 and January 1908, likewise with the aid of the recently completed NER dynamometer car. The coal consumption of each compound was carefully monitored between October 1906 and September 1907, and presumably it had been on this basis that ten more 4CCs were authorised in December 1907 to be patterned on No. 731. This had returned a slightly lower average coal consumption of 39.8lb per train mile when compared with 42.75lb by No. 730. For comparison, the best individual monthly figures from members of the Class V two-cylinder 4-4-2s were also taken and these gave an average of 42.7lb, while their overall average was 45.95lb. Once again compounding did not show up to any greatly marked advantage. W. M. Smith had died in harness after a long illness in October 1906, and the posthumous production batch of ten 4CCs in the event was never built.

Smith's successor as chief draughtsman was his former deputy, George Heppell, who had been hostile towards the 4CC project following Smith's criticism of the V class. Of strong character, he therefore refused to take any part in the 4CC design, so the third in command, Ralph Robson, had been deputed to produce the great majority of the working drawings. These were usually entered in the Gateshead drawing register in large tranches, and furthermore over an unusually lengthy period, being officially dated between March 1903 and June 1905.[8] After 1910, and the introduction of the three-cylinder simple Class Z 4-4-2s, the two NER four-cylinder compounds rapidly faded from prominence, although both were rebuilt with entirely new superheated Belpaire boilers in 1915. Interestingly, in view of associated problems with these on the Midland, both 4CC engines, and the four GCR compound 4-4-2s, retained their Smith reducing valves, located on the right-hand side of the smokebox, up to their withdrawal from service in 1933/35 and 1947 respectively.

The Great Northern Railway operated the greatest number of 4-4-2s in Britain, with a total of 116 built between 1898 and 1910, three of them originally as four-cylinder compounds, although only one of these still remained as such by 1921.

The first to appear was No. 292 in March 1905, which was based on the standard Ivatt large-boilered 4-4-2. However this had Walschaerts valve gear working the outside high-pressure slide valves, and Stephenson gear operating the inside low-pressure slide valves below the inside cylinders, which worked onto the leading coupled axle. The two sets of valve gear could be operated independently as regards cut-off. If necessary, live steam could be admitted direct to the low-pressure cylinders for starting, etc. Reflecting French practice, the boiler was designed to carry 225lb pressure, although in practice the safety valves were always set only slightly in advance of 200lb. When compared in traffic with standard No. 294 its fuel consumption was 44.0lb as against 45.3lb, a saving of only 3 per cent. Unsurprisingly, in view of No. 292's greater mechanical complexity, its lubricating oil consumption was also 15 per cent greater. The only GNR 4-4-2 never to be superheated, it was also the first of those having a wide firebox to be withdrawn, as early as January 1927, with the gap to the next being more than sixteen years.

A further Ivatt compound 4-4-2, No. 1421, was completed in August 1907, but there is no evidence of any subsequent testing, although it was paid the somewhat back-handed compliment of having been some improvement on the distinctly unreliable No. 292. A comparable four-cylinder compound 2-6-2 mixed traffic engine was also mooted at this time.[9] In December 1914, No. 1421 was superheated while still a compound, but in 1920 it was so completely rebuilt to become a standard two-cylinder 4-4-2 that was then officially considered to be virtually new. In this rejuvenated form it ran until withdrawal in August 1947, having just achieved precisely forty years' service.

[7] Smith, W. M., BP 16,310/1900 and BP 5526/1901 (relating to four-cylinder compound locomotives).

[8] Atkins, P., The Four-Cylinder Compound Atlantics of the North Eastern Railway, *Back Track*, August 1997, pp. 424–8.

[9] Brown, F. A. S., *From Stirling to Gresley*, Oxford Publishing Company, 1974, p. 98.

The third Great Northern compound 4-4-2, No. 1300, was a real oddity and might not have been either an Atlantic or even a compound. In early 1904, at Henry Ivatt's suggestion, the five largest British locomotive manufacturers were approached to submit proposals for an express locomotive capable of hauling a ten-coach train aggregating 350 tons between London and Wakefield (175½ miles) in three hours non-stop. Ivatt simply laid down the design parameters as to maximum length and weight. Between them the contractors worked up a total of fifteen schemes that encompassed a 4-4-0, several 4-4-2s, a 2-6-2, a 2-6-4 and several 4-6-0s, variously with simple or either three- or four-cylinder compound expansion.[10] These were submitted in March 1904, although none were found to be entirely satisfactory when considered in detail three months later. However, The Vulcan Foundry's four-cylinder 4-4-2 scheme was the most favoured, although subject to modifications to the boiler requested by the GNR in order to reduce the engine's total weight by 1½ tons to 70½ tons, which would also lower its rated haulage capacity to only eight vehicles. No. 1300 was duly delivered in late June 1905, but in subsequent tests between Yorkshire and London against Doncaster-built compound 4-4-2 No. 292, and a standard 'Atlantic', it proved to be the least economical in both coal and oil consumption.

In late 1914 No. 1300 was rebuilt with a superheater, but later during 1917 it was more radically reconstructed as a two-cylinder simple engine. It was withdrawn only seven years later in October 1924, after completing a much lower total mileage than other contemporary GNR standard 4-4-2s over the same period. Such was a common phenomenon over the years with singleton locomotives.

## Superheating

For some years, Midland Railway 4-4-0 No. 1044, new in 1909, showed every probability of being the last British compound to be built, in view of the then apparent absence of any further such development elsewhere in Britain. However, at this time a very significant new development, pioneered in Germany, was appearing on the scene that would be very widely and rapidly adopted to considerable effect. Like piston valves, with which it would be closely associated, superheating, or increasing the temperature of steam above that which prevailed at the point of its evaporation, had been envisaged from around the time of the appearance of the first truly *conventional* steam locomotive, the 2-2-0 *Planet* built by Robert Stephenson & Co. in 1831. The idea was to improve efficiency initially simply by reducing condensation in the cylinders. However, the necessary technologies a) to devise effective piston valves, b) achieving a satisfactory high level of superheat, and c) to overcome the associated lubrication problems, were not

collectively fully developed until the early 20th century, about seventy years later. Following earlier experiments with smokebox superheaters, Wilhelm Schmidt's fire tube superheater made its world debut in 1902 on the new Prussian State Railways Class G8 0-8-0, several thousand units of which were subsequently built in the usual KPEV fashion.

Barely two years later, in April 1904, fully appreciating the significance of this ground-breaking development, in his leisure time an enthusiastic amateur geologist, and later the author of textbooks on lubrication and meteorology, the distinctly scientifically inclined Richard Deeley on the Midland Railway sought to build forty three-cylinder compound 4-4-0s of his own design to be fitted with Schmidt superheaters. Regrettably, the MR directors balked at having to pay the then demanded patent royalty fee of £30 per engine for the superheater, while the equipment itself also then typically cost around £200–250 per unit, all of which would have been quickly recouped in terms of the resulting reduced fuel consumption. The forty engines were duly built between 1905 and 1909 using saturated steam, and would only very gradually later be rebuilt with superheated boilers between 1913 and 1926, in curious contrast to the much more rapid similar conversion between 1910 and 1914 of the ten more recently built 990 simple engines. Oddly enough, Deeley's 1904 dream somewhat improbably achieved reality twenty years later, when after a lapse of fifteen years the newly formed LMSR unexpectedly resumed construction of his compound design, with only minor modifications, in 1924, and built no fewer than 195 of these up to 1932. These were the only (conventional) British compound locomotives to be fitted with superheaters when built.

In the event, to George Churchward on the GWR would go the credit of building the first British superheated steam locomotive in May 1906, 4-6-0 No. 2901 (later named *Lady Superior*), which like two 0-6-0s that followed on the LYR later in the same year, was equipped with a German Schmidt superheater. What really put superheating on the map in Britain, however, was its early provision on some new LBSCR Class I3 4-4-2Ts in early 1908, and their subsequent participation in through running between Brighton and Rugby in trials involving non-superheated LNWR Precursor 4-4-0s on the same duties. These not only demonstrated a notable reduction in coal, but also in water consumption, which was also an important consideration in the operation of passenger tank locomotives, in view of their essentially limited capacity for both of these essential commodities.

Comparative trials by various companies over the next few years showed quite a variable range of fuel economy due to superheating, which averaged around 17 per cent:

---

[10] Groves, N., *Great Northern Locomotive History*, Vol. 3A, The Railway Correspondence & Travel Society, 1970, pp. 206–9.

London Brighton & South Coast Railway Class I3 4-4-2T, No. 22, built at Brighton Works in March 1908, when it was the first British tank locomotive to be fitted with a (Schmidt) superheater. *Former Ian Allan Archive*

## Table 21 Comparative fuel economy achieved with superheating, 1906–13

| Railway/type | Saturated engines lb coal/train mile | *Superheated* engines lb coal/train mile | Economy % |
|---|---|---|---|
| LYR 0-6-0s, c. 1906 | 73.3 | *65.0* | 11.3 |
| LBSCR I3 4-4-2Ts, 1908 | 39.4 | *34.2* | 9.9 |
| Ditto, 1907–1913 | N/A | N/A | 15.4 |
| LNWR George V 4-4-0s, 1910 | (62.3) | *45.7* | 26.7 |
| NER Z 4-4-2s | N/A | N/A | 20.0 |
| CR Dunalastair IV 4-4-0s, 1910 | 47.5 | *37.2* | 21.7 |
| GSWR 381 & 128 4-6-0s, 1911 | 54.3 | *44.6* | 17.9 |
| LBSCR 4-4-2s, H1 1906* & H2 1912* | 41.0 | *36.6* | 10.7 |
| SECR E 4-4-0s, 1913 | 38.25 | *32.0* | 16.3 |

* Six-month averages, July–December.

Some twenty-five years later the LMSR calculated decadal (1927–36) coal consumption per engine mile averages for various locomotives classes, which included former LYR 0-6-0, 0-8-0 and 2-4-2T classes that were both non-superheated and *superheated*. As with the simple *versus* compound 0-8-0s comparison made nearly thirty years earlier, in each case there was no significant *overall* difference in coal consumption per engine mile, i.e. 0-6-0 54.8/*55.1*, 0-8-0 80.7/*78.8*, and 2-4-2T 47.4/*49.0*, although the potential haulage capacity of the superheated variants would have been greater. Representatives from all six groups in question were already being withdrawn during this period, when their

collective grand total reduced from 971 to 692, although the non-superheated 2-4-2Ts and 0-6-0s did not finally become extinct until 1961 and 1962 respectively.[11]

The Midland Railway was not alone in being averse to paying patent royalties for the use of proprietary superheater designs. Over the next few years several British locomotive superintendents, sometimes in association with their senior draughtsmen, devised new superheater types of their own. Essentially these still worked on the same basic Schmidt principle of long elements being inserted in large-diameter flue tubes, but differing in detail with regard to the design and location of the header or headers, and/or the manner in which the

[11] Cox, E. S., *Chronicles of Steam*, Ian Allan 1967, pp. 116–7.

## Table 22 Summary of early fire tube superheater patents filed by British locomotive superintendents, 1908–14 only

| Rly | Patentee | Year | British Patent Numbers |
| --- | --- | --- | --- |
| GWR | G. J. Churchward* | 1908 | BP 4209, 27181 |
| MR | H. Fowler* | 1911 | BP 2445, 12884 |
| GCR | J. G. Robinson | 1911 | BP 16686, 24174, 24659, 26033, 28708 etc |
| LYR | G. Hughes | 1912 | BP 8288, 9773 |
| GNR | H. N. Gresley | 1913 | BP 4837 |
| SECR† | R. E. L. Maunsell* | 1913 | BP 3778, 19269 |
| LSWR | R. W. Urie | 1914 | BP 10782 |

* With others.

† Richard Maunsell (who was an Irishman) had applied for his patents while he was working for the Great Southern & Western Railway in Dublin, prior to moving to the SECR in late 1913.

elements were affixed to them. The patents sometimes also related to the design of associated damper or retarder devices, which were initially provided to restrain the steam temperature from rising too high by partially obstructing the gas flow through the flue tubes.

As will be seen, the GWR under George Churchward was quickest off the mark, making experiments on Saint and Star 4-6-0s that resulted in the Standard Type 3 superheater in 1909, which would then see continuous service until 1965, albeit delivering steam to the cylinders at only modest maximum temperatures of c. 500–550°F. Such Churchward deemed to be all that was necessary simply to overcome condensation in the cylinders, regardless of the enhanced cylinder efficiencies and thereby reduced coal and water consumption obtainable via higher steam temperatures. From 1910 all new GWR main line tender locomotives were provided with superheaters when built, and by the end of 1912 around 750 engines, or about one quarter of the entire GWR locomotive stock, now including some tank engines, were so equipped. Actual coal consumption figures for GWR locomotives in general, and for Churchward's locomotives in particular, are unfortunately notably lacking. By the end of 1913 only 200 LNWR locomotives were superheated, followed by the NER with 142, and Midland (excluding the ex-LTS 4-6-4T locomotives) 131.

There were curious anomalies on the LNWR. Whereas Crewe superheated many Precursor 4-4-0s, effectively converting them into George the Fifths, the corresponding Experiment 4-6-0s remained unaltered. The Whale 4-4-2Ts would also no doubt have benefitted, but always used saturated steam, as did the 19in goods 4-6-0s. On the other hand, the Class G heavy goods 0-8-0s, both those that resulted from rebuilding earlier Webb compounds, together with new builds, were in due course rebuilt into superheated G1s.

In pursuit of increased cylinder efficiency, railways other than the GWR aimed higher for the 'magic' temperature of 650°F advocated by Wilhelm Schmidt, or 'Hot Steam Willy' as he was sometimes known, but above which it was thought there could be very serious problems with valve and cylinder lubrication. Dr Schmidt's own method of restraining excessive steam temperatures, which were monitored in the cab by the driver via a pyrometer, was a servo-operated and rather complicated mechanical 'Venetian blind' arrangement mounted ahead of the upper portion of the front tubeplate and opposite the flue exits. This was not a satisfactory solution given the volcanic conditions that prevailed within a locomotive smokebox. On the LNWR at least, the Schmidt-type mechanical dampers had had one perceived virtue, however. When the fire tubes became choked with ash, drivers would temporarily 'close the blinds', thereby concentrating the draught on the lower tube bank, which thereupon became sucked clear by the exhaust.

As a more viable form of damper or retarder, John Robinson quickly devised a battery of steam jets that were centred on the individual flue exits, soon after his first superheated locomotive, Class 9J 0-6-0 No. 16, was completed at Gorton in April 1909. This had been modified with new and enlarged piston valve cylinders and equipped with a Schmidt superheater, all according to the good doctor's directions.[12] Sixteen months later, in August 1910, ten heavy suburban 4-6-2Ts and twenty heavy mineral 2-8-0s, all of new design and to be superheated, were ordered from Gorton Works. John Robinson's own initial design of superheater made its debut on the eighth 4-6-2T, GCR No. 24, in June 1911, and on his first 2-8-0, No. 966, three months later. This was only the beginning, for between 1904 and 1923 Robinson was granted no fewer than forty-five patents, nineteen of which directly related to locomotive superheaters alone.

[12] One of Dr Schmidt's recommendations appears to have been to equalise the respective free gas areas through the superheater flues and fire tubes, i.e. 50/50. However on the NER three- cylinder 'Atlantics' when introduced in 1911 this did not produce the required result. In 1914 the number of tubes was drastically reduced from 149 to only ninety to give a ratio of about 60/40, which also resulted in a 20 per cent reduction in evaporative heating surface. (The number of tubes was later substantially increased again from 1924.) Pre-1944 Swindon practice was only 30/70 (in conjunction with triple single return elements).

There was, however, a sometimes extremely strict general understanding in industrial circumstances that patentees should not derive any personal remuneration from any inventions that they might have devised in the line of duty. However, John Robinson is understood to have personally received a royalty of £50 per locomotive from the 521 2-8-0s of his design that were fitted with his superheater, which had been built by private locomotive builders during 1917–20 for the War Office. In real terms this was the monetary equivalent of about £2½ million a century later! Furthermore, around 1915 the Gorton drawing register listed numerous entries for Robinson superheater arrangements for *contractor-built* locomotives for various British colonial railways.

However, by around 1919 such paraphernalia were widely deemed to be unnecessary. Crewe had decided to remove the dampers from superheated goods locomotives after 1917, and from passenger locomotives after 1918. It was later claimed that steam temperatures as high as 800°F were attained on ex-GNR large-boilered 4-4-2s after their equipment with unusually large thirty-two-element superheater installations in LNER days. Their performance was thereby transformed. Surprisingly, some of these engines still retained their original *slide* valve cylinders,

sometimes up to their eventual withdrawal, and evidently did not encounter lubrication problems of any consequence. When the GER superheated the 'Claud Hamilton' 4-4-0s, and Robert Urie on the LSWR superheated the Drummond T9 4-4-0s and 700 class 0-6-0s, their original slide valve cylinders were also retained.

A common ploy, when superheating was becoming established and particularly when applied to already existing designs, was to increase the cylinder diameter by about 1in and correspondingly reduce the boiler working pressure by 20lb. At that time the superheater was also seen as a means of reducing boiler maintenance costs via the reduced pressure. The GSWR 4-6-0s and NBR 4-4-2s were both good examples of this early philosophy.

Experience with superheating in suburban passenger service proved somewhat varied. In late 1914 the GER built two prototype 0-6-2Ts, of which only the second was superheated. On the strength of comparisons that were made, ten production engines that were eventually built in 1921 were non-superheated. Alfred Hill later stated that there had been no difference in coal consumption and that higher maintenance costs were incurred by the superheated engine.[13] At the same discussion on the other hand, Nigel Gresley from the GNR remarked that the superheating of an

Great Eastern Railway non-superheated 0-6-2T No. 1000, constructed at Stratford Works during 1914. Companion No. 1001 was superheated and is believed to have been the last GER locomotive to be turned out painted blue. The snifting valves suggest this was actually No. 1001. (No. 1000 was not superheated until 1929.) *Former Ian Allan Archive*

[13] Hookham, J. A., Comparison between superheated and non-superheated tank locomotives, *Journal of the Institution of Locomotive Engineers*, 1922, Paper No. 126, pp. 578–633 (including discussion).

Ivatt class N1 0-6-2T in 1918 had resulted in a fuel economy of 18.7 per cent (52.3 v 64.4lb/mile). John Hookham from the North Staffordshire Railway gave results from recent trials (in 1921) with newly built Class L 0-6-2Ts, which at first appeared to favour the original non-superheated variant, represented by No. 25, only for superheated No. 72 then to show an economy of 20 per cent! Surprisingly, Hookham remarked that as regards the handsome superheated NSR Class K 4-4-2Ts that had been running since 1911, no comparisons had been possible. One would have thought, however, that a fairly accurate evaluation could have been made with reference to the non-superheated Class G 4-4-0 tender engines from which these had been directly derived, or alternatively between these and the solitary NSR superheated 4-4-0 No. 38 built in 1912, which was originally intended to be an eighth 4-4-2T.

## Compounding and superheating

The superheating of Midland Railway Compound 4-4-0 No. 1040 in 1913 enabled a unique 'four-cornered' comparison to be made as far as British practice was concerned regarding the relative economy achieved by compounding and/or superheating, which also involved the recent Deeley 990 class. The results were reported by Deeley's amiable successor, Henry Fowler.[14] The respective coal consumption in terms of lb/ton mile is summarised in the table below.

From this simple data it will be seen that superheating,

incorporated. Valid objections might have been the additional complexity involved, and the restriction of the diameter of low-pressure cylinders to a *practical* maximum of 22in, whether these were to be accommodated between or outside the main frames, taking into account the need for adequate journal dimensions and the external restrictions imposed by the British loading gauge.

The unexpected perpetuation of the Deeley Compound by the LMS apart, the compound locomotive might indeed have experienced a renaissance in superheated form in Britain. In 1923, under new LMSR auspices, Derby had proposed a rather elegant three-cylinder compound 4-6-0, and three years later actually commenced construction of a four-cylinder 4-6-2, of which there was also to be a 2-8-2 heavy goods equivalent, only to abandon the 4-6-2 in favour of the smaller and cheaper three-cylinder simple Royal Scot 4-6-0. In 1931, the complete reconstruction of a still new four-cylinder 'Lord Nelson' 4-6-0 on the Southern Railway was authorised, but not proceeded with. Meanwhile, at Swindon even the GWR had toyed with the idea of building a compound 'Castle' 4-6-0. All of these, if built, would have shortly pre-dated the outstanding developments in compounding achieved by André Chapelon in France. For its part the LNER *did* build in 1929 the celebrated Gresley experimental four-cylinder compound 4-6-2-2 No. 10,000 with Yarrow 450lb pressure water tube boiler, but with disappointing results.

### Table 23 Midland Railway 990/1000 class 4-4-0 coal consumption trials, comparing superheating and compounding, 1910–13

|  | Non-superheated (Nos 992/1032) | Superheated (Nos 998/1040) | Economy |
|---|---|---|---|
| Simple | 0.109 = 1.00 | *0.083 = 0.76* | 24% |
| Compound | 0.092 = 0.84 | *0.068 = 0.62* | 26% |
| Economy | 16% | 14% | |

whether applied either to simple or compound locomotives, conferred an economy of around 25 per cent. This transcended the also almost consistent economy of around 15 per cent achieved by compounding over simple expansion for its part. If a direct comparison is made between the extremes returned by non-superheated simple 4-4-0 No. 992 and superheated Compound No. 1040, then the combination of superheating with compounding shows a quite dramatic economy of nearly 40 per cent. By the 1920s much store was set with locomotive fuel economy, with experiments being made with feed water heaters. This makes one wonder whether compounding in conjunction with superheating should have been taken further in Britain, particularly if contemporary Swindon-style improvements in cylinder and conventional valve gear design were also

## Smokebox 'superheaters'

After only six years, by say 1912 the application of the fire tube superheater had become almost *de rigueur* (see Chronology). However, during that gestation period one or two proprietary smokebox 'superheaters' also made their appearance in the hope of increasing steam temperature, without disturbing existing 'saturated' fire tube arrangements. The first was patented by George Sisterson in 1906.[15] This received what was almost certainly its only application in September 1907 on new NER Class R 4-4-0 No. 1235, which was distinguished by its considerably extended smokebox. The apparatus consisted of two vertical headers on each side of the smokebox, which were penetrated by very short fire tubes

---

[14] Fowler, H., Superheating steam in locomotives, *Proceedings of the Institution of Civil Engineers*, 1914, pp. 77–120 (including discussion).

[15] Sisterson, G. R., BP 26,175/1906 (smokebox superheater).

GSWR superheated 4-6-0 No. 129, also fitted with Weir feed water heater. *Rail Archive Stephenson*

aligned with fire tubes in the boiler proper. A larger version was later installed and tested during the spring of 1909, when during dynamometer car tests between Newcastle and York it was assessed to have saved only 1.6lb of coal per mile. An NER test report attributed this to the equally very modest raising of the saturated steam temperature by only 25°F to 400°F. This increase was due mainly to the wire drawing of steam through the regulator, and also due to the temperature of the main steam pipe in the smokebox rather than the equipment itself, which understandably was then promptly removed.

The Phoenix smokebox 'superheater' marketed by the New Superheater Company, was trialled by several railways during 1908–14.[16] This consisted of nests of very numerous circumferential tubes within the smokebox, which therefore needed to be extended with the chimney perched at the leading end, inevitably with very ugly results.

The Furness Railway, which for some reason never embraced the fire tube superheater, made no fewer than eleven applications of the Phoenix device, the first in 1908–09 on three existing 4-4-0s and an 0-6-0, for which fuel economies of 20–25 per cent were claimed, and later had it fitted to two new 0-6-2Ts from Kitson & Co. in 1912, and to four new 0-6-0s from NBL in 1913. The Hull & Barnsley also arranged for it to be fitted to five new Class L 0-6-0s obtained from Kitson & Co. in 1911. None of these applications remained in place for very long.

Other one-off applications were made by the LBSCR and NBR to 4-4-0s, by the LNWR to a Whale 19in goods 4-6-0, and the Highland to a Castle class 4-6-0. The Brighton application, fitted to B4 4-4-0 No 59 between May 1912 and December 1915, raised the steam temperature to as high as 530°F but achieved only modest savings of 5½ per cent in coal and 8¾ per cent water. As with other applications, the device made tube cleaning and general maintenance very difficult. It was understandably loathed by the shed fitters.[17]

# Feed water heating

Feed water heaters featured only rarely in British locomotive practice before 1915, although some locomotives designed under the Drummond brothers incorporated arrangements whereby exhaust steam could be diverted into side tanks or tender tanks, as on Peter's large 0-6-0s for the GSWR, certain of Dugald's large 4-6-0s and some later M7 0-4-4Ts on the LSWR, which consequently were fitted with special feed pumps. Also on the GSWR, the second of James Manson's two superheated 4-6-0s, No. 129, was additionally equipped when new with a Weir feed water heater, which was accommodated on the top of the boiler between the smokebox and dome. This achieved only a modest *additional* saving of 3.4lb of coal per mile when compared with No. 128, in the course of the comparative trials mentioned earlier.

Also at this same period, Frederick Smith, who had recently been promoted from works manager to locomotive superintendent on the Highland Railway, devised a feed water heater, no doubt with the laudable intention of attempting to reduce the heavy coal bills faced by his somewhat impoverished company.[18] Like many others, this was intended to utilise some of the otherwise wasted heat in the exhaust steam from the cylinders, and the hot gases in the smokebox also prior to exhaust. This equipment was fitted during 1913 and 1914 to all six of the Peter Drummond Big Ben 4-4-0s, which were notoriously heavy on coal. This was almost certainly due to their over-generous boiler free gas area, which amounted to 22.3 per cent of the grate area via inappropriate 2in diameter tubes, compared to only 13.7 per cent from the more appropriate 1¾in tubes on the earlier Small Bens, both classes having the same 20.3sq ft grate area. Trials demonstrated a dramatic reduction in coal consumption of no less than 33.3 per cent (down to 45.6lb from 68.3lb per mile). Unfortunately, the estimated installation cost of £50

[16] Macaskie, S. S., BP 29,344/1909 (Phoenix smokebox superheater).

[17] Bradley, D. L. *Locomotives of the L.B.& S.C.R.,* Part 3, The Railway Correspondence & Travel Society, 1974, p. 43.

[18] Smith, F. G., BP 28,512/1912 (locomotive feed water heater).

The North Eastern Railway Class S2 4-6-0 No. 825 fitted with Stumpf Uniflow cylinders, built at Darlington Works in 1913. *Former Ian Allan Archive*

per engine had already been greatly exceeded, while the subsequent associated maintenance costs proved to be high. As with the earlier smokebox equipment mentioned, it interfered with fire tube and smokebox cleaning on shed. This was fairly swiftly removed in its entirety after Smith's sudden departure for other reasons in late September 1915. Although initiated pre-1923, the Big Ben class was superheated in early LMSR days during the 1920s.

For a brief period around 1914, HR 4-6-0 No. 141 *Ballindalloch Castle* enjoyed the extremely dubious distinction of simultaneously being fitted with both the Phoenix smokebox superheater *and* Smith's feed water heater (external version). A rare photograph of No. 141 dated August 1915 clearly reveals that the latter had already been removed by then, even before Smith's departure the following month.

## Stumpf uniflow cylinders

Two more radical attempts to improve locomotive efficiency, this time in conjunction with superheating, were made and initiated on the North Eastern Railway under Vincent Raven in the late Edwardian period. Firstly, the last of the S2 class mixed traffic 4-6-0s, No. 825, was turned out of Darlington Works in March 1913 with highly distinctive and steeply inclined cylinders that operated on the German Stumpf 'Uniflow' system. Put very simply, the cylinders were made approximately twice as long as the piston stroke, and the doubly concave and large hollow piston head was the same length as it. While live steam was admitted at each end of the cylinder by means of conventional piston valves, this was subsequently exhausted through circumferential ports

located mid-cylinder when these were uncovered by the large piston itself. As with later poppet valves on locomotives, the employment of separate inlet and exhaust ports avoided the fluctuating temperatures encountered with common steam ports that were conducive to condensation. This thereby made for greater cylinder efficiency. The downside of the Stumpf system was that it was inherently not possible to secure ideal valve events over the whole range of cut-offs.[19]

Exceptionally as far as the NER was concerned, No. 825 sported outside Walschaerts valve gear. It underwent limited comparative trials between York and Newcastle against standard S2 No. 797, which involved the NER dynamometer car, and both engines were also indicated. Loadings were varied between 223/262 and 791/800 tons, and booked average speeds were 24 and 51mph. No. 825 returned a fuel economy when averaged over these tests of around 10 per cent, i.e. 44.1 v 48.9lb/mile, and 4.12 v 4.59lb per drawbar horsepower hour. There was surprisingly very little variation regarding water consumption.

Raven felt sufficiently encouraged by this to arrange for the last engine of what would prove to be the final batch comprising twenty three-cylinder Class Z 4-4-2s, authorised in April 1914, for this also to be fitted up with the Stumpf system. The necessarily very complicated drawing for the elaborate 'monobloc' cylinder casting was executed by the same leading draughtsman, H. Spencer, who had also earlier designed the cylinders, etc, for 4-6-0 No. 825. This drawing, which unfortunately no longer survives, would have taken several weeks to produce, and was officially dated 26 October 1914 in the Darlington

[19] *Locomotives of the LNER*, Part 2B, Tender Engines, Classes B1 to B19, The Railway Correspondence & Travel Society, 1975, p. 79.

drawing register. By this time the First World War had recently begun, despite which the project nevertheless progressed, albeit very slowly and for almost the war's entire duration. As with the standard Z cylinders, the even more complex single casting for NER No. 2212 was produced in Leeds by Kitson & Co., probably at some time during 1915, before that firm had become too heavily engaged in munitions work for the government. Although the boiler scheduled for No. 2212 was built at Darlington in September 1916, the engine itself was not finally completed there until nearly two years later, in June 1918. It weighed 2½ tons more than the previous standard engines, and cost £5,261 as compared with a little over £4,000 for the others. After all this, by way of testing, according to the vehicle's log book, No. 2212 then merely undertook just three perfunctory 'maximum power' runs with the NER dynamometer car between Newcastle and York and return in early October 1918, in competition with two other recently built standard Zs, Nos 2205 and 2208. Regrettably, no results from these tests have ever been revealed, even in a contemporary detailed description of the two NER Stumpf engines that was published in *The Locomotive* in July 1919.[20]

The explosive exhaust of the Stumpf 4-6-0 in action could evidently sometimes be heard from several miles away, and by all accounts it was positively deafening if encountered when crossing within the confines of a tunnel! The engine also showed a regular propensity to burst its cylinder covers, and was converted to a standard S2 engine by the LNER as early as March 1924. By all accounts No. 2212 was more successful and it operated quite satisfactorily on normal duties in its unconventional form until November 1934. In marked contrast to the contemporary ruthless locomotive scrap and new-build policy on the LMSR, in true Gresley LNER fashion this engine was then extensively rebuilt with a newly designed boiler, and was fitted with new cylinders and with poppet valve gear. Strangely, George Heppell who as chief draughtsman would have supervised the design of both Stumpf locomotives, made no mention of either of them in his sometimes rather selective memoirs.

## Coal prices

The combined annual coal bill of the fourteen major railway companies climbed from £5.7 million in 1910 to £7.8 million in 1913, although their aggregate annual total train mileage was almost unchanged. Although the overall increase was 37 per cent, the increases experienced by the individual companies varied remarkably widely, ranging from only 17 per cent on the LBSCR to 65 per cent on the GER. Despite the absence of monetary inflation before 1914, the price of coal had long been steadily increasing over time, while by this period its quality was also perceptibly beginning to decline. The resort to cheaper coal was blamed in some quarters as having been a contributory factor in the tragic collision and subsequent train fire at remote Ais Gill in the Pennines in September 1913, although the Midland Railway denied this.[21] Taking the specific example of the North Eastern Railway, the price per ton that the NER paid for its locomotive coal, when averaged over ten year periods, were: 1881–90 37.3p, 1891–1900 45.5p, and 1901–10 51.0p, which was partly due to the fact that new mine shafts needed to be sunk progressively deeper. Average *annual* prices fluctuated, influenced by general demand during each of these periods. Whereas in 1911 the average price on the NER had been only 47.0p, after the 1912 miners' strike and the consequent wages settlement, the average price in 1913 was no less than 64.2p, i.e. up by 37 per cent within only two years. After 1914, due to wartime monetary inflation, by 1919 the actual purchase price on the NER had precisely doubled to £1.28! Such had become the price of coal by 1920 that the North Eastern, the LNWR and the Highland Railway each began to explore the feasibility of switching to oil firing, giving trials to equipment supplied by the Scarab Oil Company. It should be noted that this was well *before* the onset of the miners' strike in early 1921, which resulted in a number of applications on several other railway companies, including the Midland, Great Central, and Great Northern. The NER, however, had already discovered that comparatively speaking burning oil still cost twice as much as burning coal.[22]

There were practical and also purely theoretical limitations as to what extent the inherently extremely low *overall* thermal efficiency of the conventional steam locomotive could be improved beyond the typical maximum of only 7 per cent at the tender drawbar, which was latterly obtained under British conditions. There were, however, two simple expedients that would be the future, i.e. long-lap/long-travel (1½in/6in) piston valves working in conjunction with high-degree superheat (above 600°F), that had each independently already been established in Britain by 1914. Hitherto the two had only been combined, albeit disappointingly briefly, on GWR 4-6-0 No. 2901 when it was newly built in 1906, and two years later on four comparatively short-lived LYR 4-4-0 rebuilds. In mid-1914 Richard Maunsell decreed that these two characteristics should feature together on new 2-6-4 tank and 2-6-0 tender engines that had just begun to take shape on the drawing board at Ashford, SECR. Of these, the 2-6-0 would embrace the by then increasingly popular concept of the 'mixed traffic' tender locomotive.

[20] Three-cylinder 'Uniflow' Locomotives, North Eastern Railway, *The Locomotive*, July 1919, pp.101–3.

[21] For an in-depth discussion of this controversy, see Baughan, P. E., *North of Leeds*, Roundhouse Books, 1966, pp. 390–401.

[22] Hoole, K., *The North-Eastern Atlantics*, Roundhouse Books, 1965, pp. 25–6.

# 10
# Superheated Mixed Traffic

In 1909, Harold Holcroft, when a junior draughtsman in Swindon drawing office, obtained two months' leave in order to accompany a group of young engineers touring eastern Canada and the United States to visit various sites of general engineering interest. From the railway point of view he particularly noted the widespread use and versatility of the 2-6-0 or 'Mogul' locomotive, and he mentioned this in his subsequent report. This observation was eventually brought to the attention of George Churchward, who had rather neglected the question of building more up-to-date locomotives for secondary services on the GWR. Thus double-framed 4-4-0s, albeit with taper boilers, had continued to be built at Swindon until late 1909, paradoxically alongside the most modern 4-6-0 passenger locomotives in the land. The upshot, as Holcroft would later recall in his memoirs, was that the ever approachable Churchward actually approached *him* at his drawing board with the succinct instruction 'get me out a 2-6-0 with 5ft 8in coupled wheels, outside cylinders and the No. 4 boiler, bring in all the standard parts you can'.[1]

Holcroft also observed that what resulted, quite unconsciously, effectively amounted simply to being a tender version of the 3150 series 2-6-2 tank of 1906. Twenty 2-6-0s were immediately put in hand, with the first example emerging in June 1911, although compared with the earlier tank engine this incorporated a superheater from new, in conjunction with the newly perfected Swindon top feed arrangement for the first time. On the evidence of the Swindon drawing registers now preserved at York, it was actually Holcroft's colleague, Fred Hawksworth, not acknowledged, who really worked out the fine details on the drawing board, just as he had a little earlier with regard to the new 42XX 2-8-0 tanks for South Wales.[2] With the

[1] Holcroft, H., *Locomotive Adventure*, Ian Allan Ltd, 1962, pp. 67–71.
[2] Atkins, P., Before they were famous, *Back Track*, December 2013, pp. 725–7.

Great Western Railway mixed traffic 2-6-0 No. 4331, built at Swindon Works in 1913. The cab and framing was made 9in longer after the initial batch of 2-6-0s built in 1911. *Former Ian Allan Archive*

Great Northern Railway mixed traffic 2-6-0 No. 1630 as built in August 1912 with 4ft 8in diameter boiler. This was later replaced by one of 5ft 6in diameter in 1932. *Former Ian Allan Archive*

GWR 2-6-0 a new British locomotive genre was born, and by June 1914 sixty of the new 43XXs were already in traffic. A slightly lighter version carrying the No. 2 in place of the No. 4 boiler was subsequently also proposed, although not built. So useful did these modest engines prove that a total of 342 were ultimately constructed, the last in 1932, only four years before withdrawal of some of the older engines began. However, some parts from these were recovered for incorporation in new 5ft 8in mixed traffic 4-6-0s carrying the larger standard No. 1 boiler, a proposed ongoing conversion process, no doubt for accountancy purposes, that was curtailed after eighty Granges had thus been produced by the outbreak of war in 1939. The new 4-6-0s were better suited to higher-speed work than the 2-6-0s, having effectively been envisaged by Churchward no less than thirty-five years earlier on his far-seeing January 1901 chart! Some regarded the Granges as superior to the 6ft Halls on account of their improved cylinder design, such that their 4in reduction in coupled wheel diameter was no impediment to speed.

American-style outside-cylinder 2-6-0s were not entirely unfamiliar in Britain already, as a total of eighty had been imported from the USA during 1899–1900 by the Midland, Great Northern, and Great Central railways between them. These 'Yankees' were never popular, being regarded merely as a stopgap, and all three companies had begun to scrap them by 1910. They did, however, introduce Derby, Doncaster and Gorton to the leading pony truck, which later appeared on new locomotive designs which were first built there in 1914, 1912 and 1911 respectively.

On the Great Northern Railway in 1908, Henry Ivatt built fifteen 0-6-0s with 5ft 8in coupled wheels for fast goods work, and planned a further superheated batch of ten at the time of his retirement in 1911, which were completed by his successor, Nigel Gresley, the following year. These were sometimes used on passenger, but by their very nature were not entirely suited to high-speed work. On assuming office Gresley wasted no time in initiating an outside cylinder 2-6-0 also with 5ft 8in coupled wheels, clearly prompted by George Churchward's very recent 2-6-0s on the GWR. The first Doncaster Mogul, No. 1630, was completed in August 1912, when it made its maiden trip down to Boston in Lincolnshire. Particularly notable was the employment of readily accessible outside Walschaerts valve gear in association with two outside cylinders, for virtually the first time on a British main line locomotive.

Ten engines were built to this pattern (GNR Class H2), which were followed from 1914 by a revised design (Class H3) in which the boiler diameter was increased from 4ft 8in to 5ft 6in. Eventually all the H2 engines also received the larger boiler, although by the time the last H3s appeared in 1921, they had already been superseded by the new and rather larger Gresley three-cylinder 2-6-0s (GNR Class H4, later better known as LNER Class K3).

The next mixed traffic 2-6-0 to appear was the very businesslike Class K on the LBSCR in late 1913. This featured a new design of Belpaire boiler, but the cylinders were similar to those fitted to the recent Brighton superheated 4-4-2s and 4-6-2Ts, in conjunction with inside Stephenson valve gear. This class had the unique distinction of having its nominal designer, Lawson Billinton, build a model of one in one-sixth scale in his retirement.[3] The first two full-size engines would remain in active service for very nearly fifty

[3] Marx, K., *Lawson Billinton: A Career Cut Short*, Oakwood Press, 2007. The K 2-6-0 model is illustrated together with its maker on p. 182.

London Brighton & South Coast Railway, Class K 2-6-0 No. 346 with top feed on a local passenger train at Fratton.
*O. J. Morris/Rail Archive Stephenson*

years, until the entire class of seventeen engines was withdrawn almost at a stroke at the end of 1962. Proposals in 1919 to build a 2-6-2 tank version were rejected by the civil engineer, not surprisingly given that the maximum axle load of the tender engine was already 19¾ tons.

Richard Maunsell's Class N 2-6-0, when it made its delayed debut on the South Eastern & Chatham Railway in 1917, was notably superior technically speaking, if distinctly austere in appearance, in comparison with the recent 2-6-0s produced by Doncaster, Brighton and even Swindon. More than a century later, it is easily forgotten that this originally evolved as the mixed traffic tender version of a 2-6-4T passenger tank engine. Its boiler size had been circumscribed by the extra weight of the side tanks and their contents of the 2-6-4T, which was unfortunate given the unexpectedly short life that the tank engines as such would enjoy.

The construction of one engine of each type at Ashford was authorised in January 1915, by which time war had been declared, on which account it was not until mid-1917 that both of these were completed. Eleven more 2-6-0s were later built by the SECR, with the frames for ten of these being cut by the Midland Railway at Derby, no doubt courtesy of the Clayton connection. They concluded with a prototype three-cylinder version, No. 822, (Class N1), which was 4½in narrower over the cylinders, which had been particularly prompted by clearance problems within Mountfield Tunnel on the Hastings line. This engine incorporated Harry Holcroft's conjugated valve gear, working off the two outside valve gears in order to operate the middle piston valve, which

he had patented in 1909 while he was still at Swindon.[4]

Like the Great Central 2-8-0s, the Class N also had an alternative history in that one hundred were built by the government at Woolwich Arsenal soon after the war had ended, in order to maintain employment there. Ultimately fifty sets of parts were purchased by the Southern Railway, while some others were sold to railways in Ireland, which required them to be re-gauged to 5ft 3in.[5]

Following Dugald Drummond's sudden death on the LSWR in late 1912, his long-time works manager, Robert Urie, was appointed to succeed him in January 1913. The former chief draughtsman, J. A. Hunter, then stepped up to become works manager, and Thomas Finlayson, who had latterly been the chief designer at the North British Locomotive Company, was recruited as chief draughtsman. Like Drummond and Urie, Finlayson was also a Scotsman, and he would have been familiar with the latest Indian standard 'Mail' 4-6-0s, on which recently outside Walschaerts valve gear had become increasingly adopted with the advent of superheating. The ten Drummond four-cylinder 4-6-0s and 0-8-0s having been cancelled, Urie decided to replace these with a like number of two-cylinder 4-6-0s with 6ft diameter coupled wheels, which would utilise the already fabricated 5ft 9in diameter boiler barrels. The first entry for the new two-cylinder 4-6-0 in the Eastleigh drawing register was dated 3 March 1913, entitled 'Outline of boiler', which was followed only fifteen days later by that for what would become Urie's trademark, massive 21in by 28in outside cylinders.

Built at Eastleigh Works under the already existing H15 and K15 orders, the ten 4-6-0s entered traffic between

[4] Holcroft, H., *Locomotive Adventure*, Ian Allan Ltd, 1962, pp. 66–8, 90–6.
[5] Bradley, D. L., *The Locomotive History of the South Eastern & Chatham Railway*, Railway Correspondence & Travel Society, 2nd Edition 1980, pp. 88–91.

LSWR Urie Class H15 4-6-0 No. 483 fitted with Schmidt superheater. *Rail Archive Stephenson*

January and September 1914. Dugald Drummond had had no truck with superheating, of which Eastleigh therefore had no previous experience. Four of the new 4-6-0s were therefore turned out with Schmidt superheaters, four with the alternative Robinson pattern, and the remaining two were initially non-superheated for comparative purposes. Urie's own newly patented superheater would soon receive its first application on the solitary E14 4-6-0 rebuild No. 335, and thereafter become the standard on the LSWR during its remaining independent years.

The Urie H15 was the unwitting forerunner of some 1,350 LMS Class 5, LNER B1 and BR Standard 5 4-6-0s that would all be fitted with outside Walschaerts valve gear and 6ft–6ft 2in coupled wheels, and later built between 1934 and 1957. These had the added benefit of long-travel piston valves, which could directly be traced back to the GWR Saints via their direct 6ft successors, the 'Hall' class. Closer to home, the H15 set the pattern for the subsequent Urie N15 express passenger 4-6-0s (1918) and the smaller-wheeled S15 goods (1920), both of which were added to after 1922 by the Southern Railway, which incorporated increased valve travel and outside steam pipes, etc. Under the new regime. purely for publicity purposes *all* of the 6ft 7in N15s were then given names associated with Arthurian legend, while the final SR-built 5ft 7in S15s were not finally completed until 1936, having been delayed for several years by the prevailing adverse national economic circumstances. Even on these final engines their standard valve gear betrayed its immediate pre-1914 origins, via its short radius and long eccentric rods, in contrast to the soon to be established practice of making these respective lengths approximately equal.

The boilers of the N15s and S15s differed from those of the H15s in that their diameter at the firebox end was reduced by 3in, and the front ring was tapered, in order to save weight – about 2 tons it would appear – which seemed a little ironic considering their likewise massive construction. This included heavy cast iron dome casings, when these were normally fashioned from steel plate! The main frames of all these 4-6-0s were 1¼in thick, and almost indestructible. An interesting feature regarding these and possibly attributable to Finlayson himself, pre-NBL when a senior draughtsman with Neilson, Reid & Co, was their slight narrowing, or 'joggling', at the front end between the cylinders and the leading coupled axle. This provided more space within which to mount the outside cylinders. It was to be found on the first 4-6-0s and 0-8-0s that were built by that firm in 1902 for the Great Central Railway. It would also later feature on, among other classes, the GCR 4-4-2s and 2-8-0s, the North British Railway 4-4-2s, on the all the LSWR and Southern Railway-built H15, N15, and S15 class 4-6-0s, and not least on the 'would be' Highland Railway River class 4-6-0s of 1915.

## The Highland Railway River class – a locomotive comedy of errors, 1913–15

In mid-September 1915 the first of six large superheated 4-6-0s, which had been envisaged two years earlier and designed in early 1914, was delivered to the Highland Railway at Perth as HR No. 70 *River Ness*. However, its operation north of Perth was almost immediately vetoed

The builder's official photograph *purporting* to be Highland Railway 4-6-0 No. 70 *River Ness*. It was in reality the *final* engine, temporarily fitted with the original pattern chimney and tender coal plates, that was delivered direct to the Caledonian Railway as their No. 943 as shown on the next page.

by the HR civil engineer, Alex Newlands, and his colleague the locomotive superintendent, Frederick Smith, was dismissed only a few days later.

It has sometimes been claimed that the new engines had been designed in secret, and that when the first arrived at Perth, Newlands had had no prior knowledge of them. There was, perhaps, just a grain of truth in both statements. However, the 4-6-0s had evidently been openly referred to at the Highland Railway annual general meeting in Inverness on 1 March 1915, at which both Newlands and Smith would have been present. The following year a persistent shareholder enquired as to 'what had happened to the new engines of which we heard so much last year'. Furthermore, the new 4-6-0s' imminent appearance had been announced in *The Locomotive Magazine* for April 1915. In the event delayed, their still-awaited arrival was again briefly alluded to in *The Railway Magazine* for August 1915.

The evolution of the new 4-6-0 since mid-1913, when quotations were first sought to obtain four locomotives in time to handle the 1914 heavy summer traffic was, however, distinctly unusual. Although no order actually ensued at this stage, this did result in an unidentified North British Locomotive Company draughtsman unwittingly producing a flawed diagram, dated 26 August 1913, which showed an approximate weight in *working order* of 66 tons (which nearly twenty years later would be quoted as the *empty* weight), together with a coupled axle load of 16 tons. Owing to the limited drawing office resources at Inverness, five months later in January 1914, it was arranged that the North British *Railway* at Cowlairs would prepare the working drawings and draft the specification for the new 4-6-0 for the Highland Railway at a maximum cost of £200. For its guidance the HR sent on to Cowlairs the NBL diagram, a copy of which was discovered in 1976, part of which is reproduced below.

The fateful NBL project diagram for the Smith 4-6-0 dated August 1913, that would unwittingly have dramatic consequences two years later. *Author's collection*

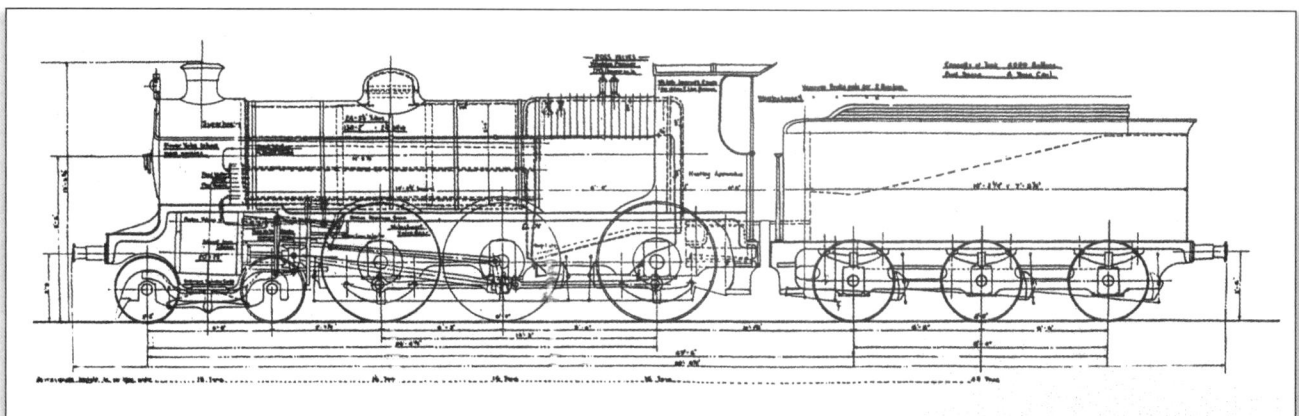

Archibald Campbell, a leading draughtsman at Cowlairs, who had earlier been closely involved in the design of the NBR 4-4-2s, was deputed to undertake the detailed design of the new Highland Railway 4-6-0. He worked in several boiler and cylinder details that were taken from the 'Atlantics', and also copied in their drop grate. Campbell also increased the cylinder diameter from 20in to 21in, copying the barrels of the new cylinders recently designed to be fitted to the NB 4-4-2s when they were superheated, and he likewise now pitched their centres 1½in further apart.

It was after the declaration of war in early August 1914, with the previously unanticipated realisation of the HR's now national strategic importance with regard to the movement of naval personnel and supplies to and from Thurso, that quotations were sought, this time for *six* superheated 4-6-0s. (These would not themselves work any further north than Inverness). Once again NBL's quotation was rejected, this time in favour of that from R & W Hawthorn Leslie & Co. in Newcastle on Tyne. As the builders later were not held to be at fault, at that stage they must have accurately calculated the working weight to be around 70+ tons, which as per normal procedure was not indicated in the printed specification. For his part Alex Newlands must hitherto have been content with NBL's apparent 'estimate' of only 66 tons. In view of the

later HR Clan class 4-6-0 (1919) it is highly probable that the *intended* axle load had been no more than 16 tons, i.e. only 1 ton more than that of the 15 tons of the Castle class 4-6-0s introduced back in 1900.

Although delivery had been contracted to begin in May 1915, due to wartime exigencies the first engine was not in fact despatched from Newcastle until 14 September, and so would probably have arrived at Perth about two days later. *River Ness* is definitely known to have progressed north from there as far as Aviemore, on what could have been its first journey under its own steam. It was rumoured locally to have scraped the platform edge at Dunkeld, only a few miles north of Perth, which possibly resulted from the enlargement of the cylinders.[6] However, it was the weight, 'found to be considerably heavier than the directors had anticipated', that was the problem. This stood at about 71½ tons in association with an axle load of 17¼ tons, the anticipation of which Newlands cannot have been advised of a year earlier, possibly on account of the urgent situation. This would nevertheless have been a regrettable error on Smith's part.[7] A slightly higher axle load of 17.8 tons had already been accepted for several years on the Big Ben 4-4-0s, whose *weight per foot run* of coupled wheelbase was also greater, yet later at the 1916 Highland Railway AGM weight in relation to wheelbase

The last River photographed as CR No. 943 in Newcastle on 13 January 1916. It was despatched north eight days later in *dark* blue CR livery. The builders commissioned three fine ¹⁄₁₆ scale models of this engine, two of which are known still to exist, held in museums at York and in Newcastle.

[6] J. R. Morrison to author, December 1984. The HR loading gauge was actually slightly amended under Newlands' signature in mid-November 1915, i.e. *after* the sale of the Rivers to the Caledonian Railway had been agreed, such that the River 4-6-0s would then have cleared this!

[7] On the later LMSR Northern Division engine diagram made for this class, significantly the difference between the empty and working weights was 5.8 tons!

was stated to have been the issue. On 24 September Smith was called to appear before a hastily convened special meeting of the HR Board at Perth, at which his immediate resignation was requested. Smith evidently later never spoke of the issue, nor did he ever return to railway work, although his wife was heard to remark bitterly that if he had had a 'Mac' in front of his name 'all would have been well'.[8]

In the aftermath, all six Rivers were sold on to the Caledonian Railway, becoming CR Nos. 938–43. Shorter chimneys were fitted, and the drop grates and Smith feed water heaters, already installed, were both removed by the CR so as to absolve it from having to pay patent royalties on the latter.[9] Although on the Highland Railway the new 4-6-0s had been intended to be passenger locomotives, their 6ft diameter coupled wheels would have ranked them as mixed traffic engines on most other lines. After briefly employing them on express passenger duties the CR very soon deployed them on fast goods workings, which in wartime took priority.

In early LMSR days, in 1928, after thirteen years the Rivers belatedly began to work between Perth and Inverness, after the Bridge Stress Committee data revealed their unusually low hammer blow, which was equivalent to only 2.5 tons at 5rps (64mph).[10] The rationale behind this low figure, although sometimes ascribed to Smith's genius in order to mitigate the heavier axleload, is unclear, but this would not have been particularly necessary if the original *intended* axle load had been only 16 tons. The balancing calculations were more likely to have been made at Cowlairs, quite possibly with the earlier conflict with the civil engineer over the NBR 4-4-2s still in mind. Below the boiler, the Urie LSWR H15, both mechanically and dimensionally, was remarkably similar to the River class.[11] Both had 6ft diameter coupled wheels but, as befitting its massive physique, at any given line or rotational speed the Urie engine would have delivered almost three times the hammer blow of the River (for example 7.25 tons at 5rps) on top of a static axle load of 19.8 tons.

The Rivers would appear to have influenced latter-day locomotive design at St Rollox. Outside Walschaerts valve gear was later employed on the four extremely unsuccessful William Pickersgill three-cylinder 956 class 4-6-0s built in 1921, which were originally designed with a complicated unpatented conjugated motion to work the middle piston valve. It was also employed in association with slide valves on the eight rather odd non-superheated 191 class 4-6-0s

for the Callander & Oban line, which were commenced at St Rollox in 1920 but completed by NBL in 1922.

Walschaerts valve gear would likewise have graced a powerful 2-6-0 with 5ft 6in coupled wheels proposed in December 1920, with a 5ft 9in diameter Belpaire boiler and estimated to weigh 72 tons, which was allegedly inspired by the very recent Gresley three-cylinder 2-6-0s of similar weight on the GNR. The CR engine would almost certainly have been fitted with River class cylinders, the patterns for which had earlier been purchased from Hawthorn Leslie & Co. Production of the working drawings did not actually begin, however, until mid-1921, very shortly after which the fate of the Caledonian Railway was sealed by an Act of Parliament. This process nevertheless proceeded slowly until the end of June 1923, after which the CR was finally legally absorbed six months late by the already existing LMSR, and all work on the proposed 2-6-0 abruptly then ceased. However, via its 21in diameter cylinders, it did at least provide the starting point for the Hughes 'Crab' 2-6-0 designed under George Hughes for the LMSR in its earliest years and introduced in 1926.[12]

The proposed Pickersgill 2-6-0, with 5ft 6in coupled wheels, would have been a far cry from J. F. McIntosh's 179 class 5ft 9in inside-cylinder 4-6-0s built in 1913–14 essentially for fast goods work, for which one can also read mixed traffic when it came to working excursions and football specials, etc. These had been simply superheated versions of his 908 *passenger* engines of 1906. As with the 918 5ft 4-6-0 goods engines also built back then, the very similar boiler proportions of all three classes, which were directly derived from that for the 6ft 6in 903 class with its larger firebox, were far from ideal. They had short and very shallow fireboxes, and free gas areas through the tube bank that amounted to around 20 per cent of their small 21 sq ft grates. As a direct consequence the two non-superheated classes are reputed to have burned coal at a rate of 90 to 100lb per mile, although a superheated 179 when newly out of works could run on about half of that. Although starting at 48lb this would, however, gradually increase to 70lb/mile as the single solid piston valve rings became worn over the course of eighteen months, when the next shopping would be coming due. A photograph of one of these engines on shed shows its tender piled well beyond its nominal capacity with large coal. The LMSR would later equip one of this class with a Worthington Simpson feed water heater and pump in an attempt to reduce coal consumption.

---

[8] For a more detailed account by the present writer of this locomotive contretemps, see The Highland Railway 'River' class affair, September 1915, in *LMS Journal* No. 7, 2004, pp. 52–69.

[9] G. R. M. Miller to author, May 1978 (personal recollection).

[10] It was common practice to balance 66% of the weight of the reciprocating masses on express passenger locomotives, making for high hammer blow. In the British Railways Standard Class 7 'Britannia' 4-6-2 (1951) the proportion was only 40%, giving a hammer blow of merely 2.6 tons at 5rps (66mph).

[11] At a meeting in London, presumably in 1914, Smith told Urie that he had designed a similar engine for the HR. (J. C. Urie to author, July 1973). However, the latter's format was already established by September 1913, i.e. four months before Urie's engine entered traffic on the LSWR in January 1914.

[12] Cox, E. S., *Locomotive Panorama*, Vol. 1, Ian Allan, 1965, pp. 31–8.

The CR 179 class and the exactly contemporary Great Central 1A class were each the sixth class of 4-6-0 to be introduced since 1902 by their respective companies, yet each had three more 4-6-0s of their *own* design still to go. The lead engine of the Great Central class also bore a distinctly Scottish name, *Glenalmond*, that being the Perthshire residence of the company chairman, Sir Alexander Henderson (who was later raised to the peerage in 1916 as Lord Faringdon). Intended for express goods working, at 5ft 7in these engines were simply smaller-wheeled versions of the Sir Sam Fay class. They could be said to have been to the Sir Sams what the 19in goods on the LNWR were to the Whale Experiments. The author's father, who regularly observed, and heard the Glenalmonds at work in the Nottingham area between the wars, used to recall that they were extremely noisy engines.

In 1914 the Great Eastern Railway proposed a 5ft 8in version of its 1500 class passenger 4-6-0 but with a slightly larger boiler and having a 17-ton axle load. Although for many years it had been almost exclusively an inside-cylinder line, this was followed by a further closely related proposal that would have had 20in by 28in *outside* cylinders, with Walschaerts valve gear, likewise driving onto the leading coupled axle. (Equally puzzling was a later proposal made at Stratford in 1917 for a direct 4-6-2 tank engine version of the 1500 class 4-6-0.)

CR 179 class 4-6-0 No. 188 on shed at Perth, c. 1922. *W. H. Whitworth/ Rail Archive Stephenson*

GCR 1A class 4-6-0, No. 4 *Glenalmond,* at Nottingham Victoria Station in 1923. *T. G. Hepburn/Rail Archive Stephenson*

In July 1918 Gorton Works turned out, still in sequence, an *outside*-cylinder equivalent of the Glenalmond, simply by altering one of a batch of new large-boilered 8M 2-8-0s, No. 416, during construction into a 4-6-0, (Class 8N). The initial diagram had been prepared in December 1917 and the requisite new frame drawing was dated three weeks later. Curiously, the coupled wheel diameter was increased from the established Gorton standard of 5ft 7in by just 1in to 5ft 8in, although such minutia was actually highly significant.[13] This was at the time when the ARLE was inviting its members to submit proposals for future national standard locomotives, and 5ft 8in was the stipulated mixed traffic coupled wheel diameter, this already being the case at both Swindon and Doncaster. Indeed, in the railway press the Great Central 8M 2-8-0, which had already been ordered back in 1916, together with the new hurriedly configured 4-6-0, were both unusually described as Standard.[14] There were also the 4-6-0 and 0-8-0 parallels that had been similarly evolved almost simultaneously and quite independently at Darlington on the North Eastern Railway, as previously described.

The 8N was all round the best proportioned, not least boiler-wise, of the *nine* different 4-6-0 classes that were ultimately produced by and for the Great Central Railway over the course of almost twenty years. However, only two more of these were built, with small side window cabs, in 1921. Almost inexplicably, the GCR then moved on instead to produce a more elaborate *four*-cylinder version of the inside-cylinder Glenalmond with its very lengthy and poorly proportioned boiler. This design was distinctly heavier and more complicated, and therefore would have been more expensive to build, operate and maintain than the two-cylinder 9N. Nevertheless, no fewer than thirty-eight of these albeit extremely handsome four-cylinder 4-6-0s were then constructed by Gorton Works, and by the Vulcan Foundry and Beyer, Peacock & Co., the last few being turned out under LNER auspices in 1924.

In simpler two-cylinder form the superheated mixed traffic locomotive really came into its own in Britain after the end of the First World War. Generally enhanced by long-travel valve gear, it chimed very well with the greatly changed operating conditions brought about by the institution of the long-awaited eight-hour working day for footplate men in 1919. Between 1923 and 1937 the Big Four grouped railways between them built a total of around 350 2-6-0s having 5ft 6in–5ft 8in diameter coupled wheels. The precursor of them all could be said to have been the pioneer mixed traffic Mogul, GWR No. 4301, back in 1911. Churchward broke new ground again when, in May 1919 only a few months after the Armistice had been signed, he completed his prototype heavy mixed traffic 2-8-0 No. 4700 with 5ft 8in coupled wheels. This was temporarily fitted with the Standard No. 1 taper boiler, and in this short-lived form was therefore almost entirely constructed out of pre-1914 designed major components. On account of their restricted route availability, only eight more of these locomotives were later added during 1922–23. These were provided from new with the new larger Standard No. 7 boiler, which Fred Hawksworth stated had been expressly designed for the class, and which was retrofitted to No. 4700 in May 1921. Replacement No. 7 boilers were later built at Swindon in the 1950s. However, this time no other British railway followed suit, nor ever appears to have considered building any *eight*-coupled mixed traffic locomotives thereafter. By 1938 6ft two-cylinder 4-6-0s were in the ascendant on both the GWR and LMSR, while no new 2-6-0s at all were built by any of the four railway companies during 1939–45.

After the Second World War had ended no fewer than 550 new 2-6-0s were then built between 1946 and 1957, but this time with coupled wheels of only 5ft to 5ft 3in diameter. Most of these had originally been conceived by the LMSR (H. G. Ivatt Classes 2 and 4) and LNER (Peppercorn Class K1) simply as long overdue replacements for outdated inside-cylinder 0-6-0 goods engines. British 2-6-0s and 4-6-0s (of all kinds) would both achieve their all-time numerical peak totals during 1957, when the final examples of each type were built by British Railways. Harold Holcroft would nonetheless outlive them all, although dying only fourteen years later in 1971, which also happened to be just sixty years after the emergence of GWR No. 4301.

[13] Atkins, P., Much ado about nothing (British national standard locomotive proposals, 1917–18), *Back Track*, August 2008, pp. 461–7.

[14] Standardisation of Locomotives: Great Central Railway, *The Railway Engineer*, March 1918, pp. 46–8.

The shape of further things to come? Great Western Railway mixed traffic 2-8-0 No. 4700 as originally built at Swindon Works in early 1919, when it was initially fitted with the Standard No. 1 boiler. *Former Ian Allan Archive*

# 11

# Tank Locomotives

In 1913 tank locomotives accounted for 40 per cent of the national locomotive stock, a high proportion that was largely due to British geography, and which would decline only very slightly over the course of the next forty years. Tank locomotives were employed almost exclusively on many smaller railways, particularly in South Wales, and by the London Tilbury & Southend Railway, for example. As regards the fourteen major railways, the proportion of tank engines ranged from as low as 7 per cent on the Glasgow & South Western Railway, to as high as 73 per cent on the London Brighton & South Coast. Quite coincidentally, however, the 'last word' in new-build locomotive development on both of these two railways would be very imposing 4-6-4 tank engines having unusually large 22in diameter outside cylinders with outside Walschaerts valve gear, which were introduced in 1922 and 1914 respectively.

At 55 per cent, tank engines were surprisingly dominant even on the far-flung Great Western Railway. In fact, just over one third of its total locomotive stock in 1913 was comprised simply of 0-6-0Ts, which at that time were almost entirely of the saddle tank variety, and there

LBSCR Class L 4-6-4T No. 328 passes Merstham with the down 'Southern Belle' Pullman in 1923.
*O. J. Morris/Rail Archive Stephenson*

would be many hundreds more 0-6-0 tanks still to be built by and for the GWR, but of the highly distinctive pannier type instead, in order to accommodate the now standard Belpaire firebox. 0-6-0 tank engines in general, which were the second most numerous British locomotive type after the 0-6-0 tender engine, were employed by almost every railway company, large and small, to some degree, (including the Great North of Scotland). In December 1914 the fourteen major companies alone collectively owned around 3,400 0-6-0 tank engines, but relatively few of these had been built since 1900. Furthermore, none of these had been built by the London & North Western, Lancashire & Yorkshire nor London & South Western railways since before that date.

At the other extreme, the Edwardian era was also characterised by the production of a number of new large tank engine designs, many of which were equipped with superheaters, and which sometimes introduced entirely new wheel arrangements, such as the 4-6-4T mentioned above, to the then ever-changing British railway scene.

## Passenger tanks

At the dawn of the Edwardian era, suburban and branch line passenger workings were dominated by four-coupled tank engines, 0-4-4Ts and 2-4-2Ts in particular. After 1900 the 0-4-4T was favoured over the 2-4-2T roughly two to one by the major railways, although the Great Eastern built both. The heaviest exponents of each type were the Dugald Drummond Class M7 0-4-4Ts on the London & South Western, with 105 examples built between 1897

and 1911, and the final series of 2-4-2Ts on the Lancashire & Yorkshire, for which no fewer than 330 2-4-2Ts were built in total by John Aspinall and George Hughes between 1889 and 1911.

Compared with the LYR 2-4-2Ts, the Drummond M7s incorporated only relatively minor successive variations, but interestingly whereas the LSWR at the time quoted their full weight at only 54½ tons, the Southern Railway later gave this as 60 tons. This significant anomaly must have come to light many years later following a systematic weighing of representatives of the larger ex-LSWR locomotive classes in 1924, which had been ordered by the CME.[1] The lower weight had been stated at the Board of Trade enquiry following the derailment near Tavistock in March 1898 of a newly built M7. The LSWR was criticised for employing heavy front-coupled tank engines on fast passenger duties. Drummond then (temporarily) reduced their boiler pressure from 175 to only 150lb, possibly by way of an attempt to restrain their over enthusiastic operation.

This derailment made for an interesting prequel to an accident that befell one of the superheated batch of LYR 2-4-2Ts fourteen years later. These had immediately been put onto express workings, even taking over from the recent big four-cylinder 4-6-0s despite, like the Drummond M7, also having originally been designed for purely local passenger work. This ultimate batch, built at Horwich Works in 1911, weighed 66¼ tons, 10¼ tons more than the original 1889 design, although of this 3½ tons was accounted for by a heavy casting set between the frames at the back end to counterbalance the now much heavier front end due to the superheater headers and much larger

---

[1] Holcroft, H., *Locomotive Adventure*, Ian Allan Ltd, 1962, pp. 124–5.

**LYR 2-4-2T No. 18, the first of the final twenty engines built in 1910–11 and superheated from new.**
*P. F. Cooke/Rail Archive Stephenson*

cylinders, which over time had become enlarged from 17½ to 20½in diameter. In July 1912 one of these new 2-4-2Ts derailed at 60mph on a curve that was subject to a 30mph speed restriction near Hebden Bridge, a bizarre consequence of which was the ejection onto the ballast of an occupied coffin bound for Harrogate! This incident also attracted unfavourable comment from the Board of Trade inspector, and the 2-4-2Ts in question were promptly transferred to other less spectacular duties.

Remarkably, during 1913 the 330 LYR 2-4-2Ts worked 59 per cent of the railway's passenger mileage, or an average of 23,559 miles per engine. Some of the Drummond M7s operated for sixty-five years overall, as did the pioneer LYR 2-4-2T No. 1008 (now preserved). It is a curious fact, however, that although some 1910-built saturated steam 2-4-2Ts were being rebuilt with superheaters until as late as 1925, the very last of the LYR radial tank engines to be built, No. 627 new in November 1911, was also the first of the 330 to be scrapped, in August 1927, after less than sixteen years.

After 1900 new 0-4-4T classes appeared on the SECR, the GSWR, the North Staffordshire Railway, and the NBR, although the latter did not make such extensive use of the type as did the Caledonian, which continued to build what was essentially the same 1895 design through to 1922. The SECR had later also proposed a second rather larger 0-4-4T design in 1910, which was not built. A solitary 0-4-4T was built for the Wirral Railway by Beyer, Peacock & Co. in 1914, whose design, in the interests of lower cost and local standardisation, was very closely related to that of the two goods 0-6-4Ts that it had supplied to the WR back in 1900.

Following a prototype built in 1900, the GWR at Swindon built thirty inside-cylinder 2-4-2Ts at Swindon during 1902–03 for suburban passenger work. These were early exponents on the GWR of piston valves, which unusually were set one above the other on the vertical centre line. The first twenty had domeless, parallel Belpaire boilers that very quickly began to be replaced by taper boilers from 1905, while the last ten came out new with coned boilers. Unlike their LYR counterparts, which had reversible water scoops, these engines had separate scoops for working in each direction.

The GER built no fewer than 166 2-4-2Ts of three classes at Stratford Works between 1901 and 1912, of which the most diminutive were twelve built in 1909–10 for light branch line, rather than suburban, service. These were distinguished by conspicuously large cabs with side windows, which resulted in them being termed 'Crystal Palaces'.

The 4-4-2 tank engine, with outside cylinders, first appeared on the London Tilbury & Southend Railway in 1880, designed according to a specification drafted by William Adams, late of the Great Eastern Railway, and now on the London & South Western Railway, for which Beyer, Peacock & Co. built the first 4-4-2Ts in 1882. Before 1901 the type was also established on the Taff Vale, where it was the first with inside cylinders in 1888, and Great Northern Railways, also having inside cylinders. During the Edwardian period the 4-4-2T would be

The contemporary 4-4-2 tank engines of the LNWR and GWR were each directly derived from corresponding 4-4-0 tender locomotives, but thereafter all semblance ended. They very neatly demonstrated the very wide divergence in the respective standard practices of the two companies, i.e. inside *versus* outside cylinders, and round-topped *versus* tapered Belpaire boilers.

*Below:* LNWR Precursor tank No. 762. At this period the company began to apply its insignia to the side tanks of its 4-4-2 and 4-6-2 passenger, and 0-8-2 goods tank locomotives. *Former Ian Allan Archive*

adopted by several other railways, on which it was almost invariably elegant and well proportioned. One of the first examples was on the Great Central under John Robinson had adopted the type (Class 9K), in succession to his predecessors' 2-4-2Ts. The original thirty GC 4-4-2Ts had handsome antecedents in Ireland from when Robinson had previously served the Waterford Limerick & Western Railway. The former were added to in 1907 by a further series of twelve engines (Class 9L) which had slightly increased water capacity.

The LBSCR was the most prolific in this particular field, with Brighton Works turning out a total of sixty-two 4-4-2Ts between 1907 and 1913. These were comprised of no fewer than four distinct classes, classes I1 to I4, that were introduced within only sixteen months between June 1907 and September 1908. The initial I1s, together with the similar I2 and I4 classes, all with 5ft 6in diameter coupled wheels, were a distinct disappointment, but not so the 'express' I3s with 6ft 9in coupled wheels, several of which ranked as the first superheated locomotives in southern England. Experimental through running between Brighton and Rugby during October 1909, which also involved a non-superheated LNWR Precursor 4-4-0, No. 7 *Titan*, dramatically demonstrated the advantages of superheating. Two superheated I3s returned coal consumptions of 27.4 and 28.1lb/mile, against 41.2lb by the LNWR tender engine, and water consumptions of 22.4 and 22.7 gallons/mile as against 36.6 gallons. It was no wonder that Crewe wasted no time in designing a superheated version of the Precursor, of which No. 2663 *George the Fifth*, duly appeared only nine months later.

The 4-4-2Ts of the GWR and the LNWR made an interesting contrast. Both were direct developments of earlier 4-4-0 tender locomotives of the respective County and Precursor classes. The GWR engines were, however, fitted with the slightly smaller Standard No. 2 boiler, although one engine was turned out new in October 1906 with the larger No. 4. It ran in this condition for only 6,000 miles over three months before it was converted to standard, owing to overheating of the coupled axleboxes due to the increased loading. Both of Churchward's large outside-cylinder four-coupled classes enjoyed relatively short lives when compared to the others that had been built alongside them, being retired in the early 1930s, together with many double-framed 4-4-0s of similar vintage. In 1913 a prototype small 4-4-2T, No. 4600, schemed and largely designed by Fred Hawksworth, originally with the unique axle spacing of 7ft + 7ft + 7ft + 7ft, was built as a potential replacement for the ageing Metro 2-4-0Ts. No more followed, and it was withdrawn earlier still, in 1925, although the last 2-4-0T would not be retired until 1949.

The LNWR 4-4-2Ts differed from the corresponding 4-4-0 tender engines in having smaller (6ft 3in) coupled wheels. With nearly 4 tons of water in *each* side tank, quite apart from the weight of the tanks themselves, one suspects that these engines were probably rather heavier than was officially admitted, because the weight diagrams showed only a ¾-ton increase in axle load over the 19 tons of the Precursor 4-4-0. Indeed, an inside-cylinder 2-6-4T with the much reduced axle load of 15 tons was later proposed in October 1909, which was later redrafted as a 4-6-2T.[2] Unlike many of the Precursor 4-4-0s, the LNWR 4-4-2Ts were never superheated, although there is evidence that in 1910 some of these were equipped with a form of feed water heating apparatus together with feed pumps.

[2] Reed, B., *Crewe Locomotive Works and its Men*, David & Charles, 1982, 2-6-4T diagram on p. 135.

*Below:* GWR County tank No. 2241, the first of the superheated final series built at Swindon Works in 1912. *Former Ian Allan Archive*

LTSR 4-4-2T No. 80, *Thundersley* (1909). *Rail Archive Stephenson*

## Adams family 4-4-2Ts:

In 1912 the Midland Railway inherited seventy 4-4-2Ts following its absorption of the London Tilbury & Southend Railway. Steadily developed under Thomas Whitelegg from the original William Adams specification drafted at Stratford in 1879, the largest and most recent of these were four that had been built by Robert Stephenson & Co. in 1909. These were similar to some 1905 rebuilds with larger boilers of then only eight-year-old 4-4-2Ts, in order to keep pace with steadily increasing train loadings. These had increased from 67 tons tare in 1881, to 143 tons in 1891, and to no less than 304 tons by 1910. With these, however, the limit had finally been reached, although it is understood that their true weight had been concealed from the Great Eastern Railway, over whose lines they were obliged to operate. Each LTSR locomotive was individually named after locations on the railway. However, this policy sometimes resulted in the bewilderment of elderly passengers, who mistook them to be actual

MGNJR 4-4-2T No. 9 (1904). *Former Ian Allan Archive*

NSR 4-4-2T No. 55 (1911) at Manchester, London Road Station. *P. F. Cooke/Rail Archive Stephenson*

destinations! (Similar misunderstandings likewise arose on the LBSCR, which until 1905 had similarly topographically named its passenger tank locomotives.) The great majority of the Tilbury 4-4-2Ts were also accorded an elaborate lined green livery that, like the names, would very rapidly disappear under crimson lake paint after 1912.

Repeat orders for 4-4-2Ts of this final Tilbury design, with 6ft 6in diameter coupled wheels and surprisingly still not superheated, were built for the LMSR until as late as 1930. The last British 4-4-2Ts to be built, these could directly trace their origins back to the first built just fifty

years earlier, while several Edwardian 4-4-2Ts on three other small railways also had close Adams associations.

William Adams' son John, while locomotive superintendent of the North Staffordshire Railway between 1902 and 1915, built seven superheated inside-cylinder 4-4-2Ts at Stoke during 1911–12 that, their Belpaire fireboxes apart, quite remarkably resembled the Brighton I3 class. They were, however, directly derived from the four non-superheated NSR Class G 4-4-0s that had been built in 1910 at Stoke. One NSR 4-4-2T was stationed at Derby and worked to Birmingham, while the others at

FR 4-4-2T No. 38 (1915). *Former Ian Allan Archive*

Stoke appeared daily at Manchester, London Road. There had been an earlier proposal for a smaller 4-4-2T, which would have been of very similar proportions to its counterparts on the Great Central. Also, William Pettigrew, the contemporary locomotive superintendent of the Furness Railway, had previously enjoyed a long association with William Adams as his works manager at both Stratford and Nine Elms. His final design for the FR was an attractive small inside-cylinder 4-4-2T, of which six were delivered during 1915–16 to pre-war designs.

More obscurely, between 1904 and 1910, the Midland & Great Northern Joint Railway at its Melton Constable works in rural Norfolk, turned out three outside-cylinder 4-4-2Ts. These were virtually tank versions of four 4-4-0 tender engines with unmistakeable Adams features, built for its predecessor, the Lynn & Fakenham Railway, by Beyer, Peacock & Co. back in 1882. Then that builder also had on order the first batch of 4-4-2Ts for the LSWR placed by William Adams. Prior to the arrival of Dugald Drummond at Nine Elms in 1897, for many years there had been a particularly strong association between the Manchester builder and the London & South Western Railway, which stretched back to the time of Joseph Beattie.

The GNR continued to build 4-4-2Ts, fitted with condensing gear for tunnel working, until 1907. Further north, in 1911 the North British Railway ordered thirty 4-4-2Ts from the Yorkshire Engine Company. Dimensionally speaking, these were no advance on the very recent NBR 0-4-4Ts built in 1909, except that they had the advantage of having increased coal and water capacity. A superheated development of these was later introduced in 1915, which together with further engines built in 1921, were all constructed by the North British Locomotive Company.

A logical development of the 4-4-2T was the rare 4-4-4T, pioneered in 1896 on the Wirral Railway, which took delivery of two more modified engines in 1903. At one point the GWR had contemplated extending its County 4-4-2T with a larger bunker supported by a trailing bogie. The only *major* British railway ever to adopt the type was the North Eastern in 1913, with its new three-cylinder superheated Class D. A total of forty-five of these, some of them deferred by the war, were ultimately built until 1922. A particular disadvantage of this wheel arrangement was the low proportion of adhesive weight in relation to total weight, which in this instance was only 45 per cent, even with full side tanks. The entire class was therefore later rebuilt by the LNER to become 4-6-2T between 1931 and 1936, whereby the adhesive weight was increased from 40 to 52¼ tons, or now 60 per cent of the now slightly reduced total engine weight.

Indeed, even by 1914 the four-coupled passenger tank locomotive was visibly giving way to six-coupled locomotives having increased adhesive weight in relation to their size. The 0-6-2T was well established, particularly in South Wales, by 1900 as the tank engine equivalent of the 0-6-0 goods engine, which soon after began to be built with larger diameter coupled wheels. A good example was

NER Class D three-cylinder 4-4-4T No. 1326, when brand new in 1922. *W. H. Whitworth/Rail Archive Stephenson*

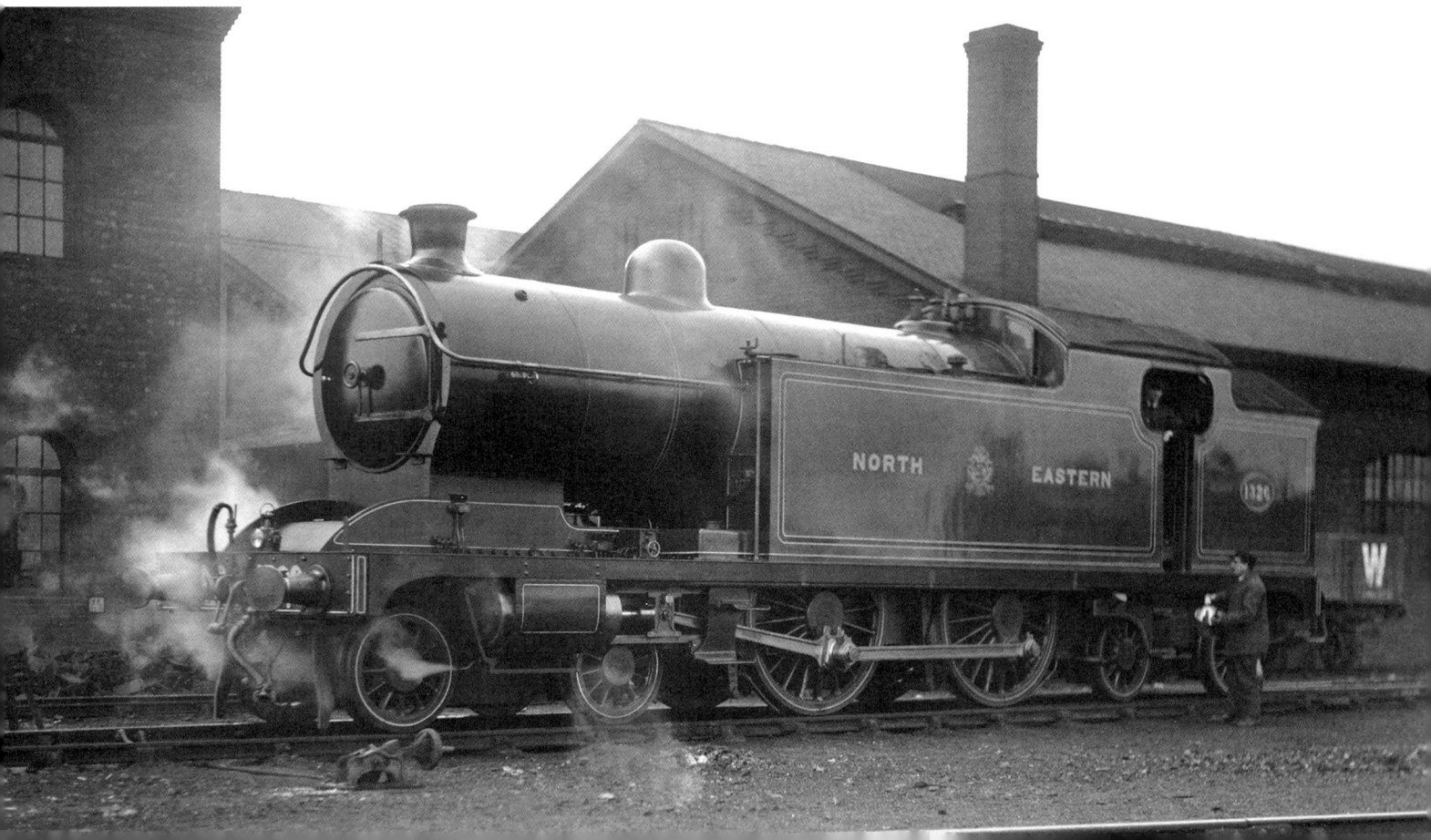

LBSCR Class E5 0-6-2T No. 573 *Nutbourne* (1903) in Stroudley yellow livery. *Former Ian Allan Archive*

GNR Class N1 0-6-2T No. 1569 (1910) with condensing gear, in green livery. *Former Ian Allan Archive*

the thirty-strong 5ft 6in E5 class on the LBSCR built between 1902 and 1904, while in 1907 the GNR began to build 0-6-2Ts with 5ft 8in wheels in immediate succession to its final batch of 4-4-2Ts, following the construction of a single prototype with larger side tanks a few months earlier. More of these engines were then built in regular batches until 1912, bringing the total of fifty-six, of which just four were not equipped with condensing apparatus for service away from London.

The GER completed two 0-6-2Ts of new design at Stratford Works in late 1914. An unusual feature of this design, of which many more were built under LNER auspices, was the employment of inside Walschaerts valve gear.

The next step was the passenger 0-6-4T, and at the time it must have appeared uncharacteristically bold when the Midland Railway rapidly turned out forty of these with 5ft 7in coupled wheels, at the rate of one per week during April to December 1907. In fact, these were as nothing when compared to what had been variously contemplated at Derby since 1903, when shortly before Samuel Johnson had retired ten inside-cylinder 4-4-4Ts had been authorised, only for these to be cancelled by Richard Deeley. There followed a surprising number of remarkably ambitious, some bordering on the bizarre, alternative proposals for *six*-coupled passenger tank engines of various wheel arrangements. These were prepared over a period of two years by the recently arrived and highly experienced Scottish leading draughtsman James Anderson.[3] The more prominent of these, with their *estimated* weights were:

[3] Atkins, P., The evolution of the 'flatirons', *Midland Record* No. 9, 1999, pp. 4–20.

Midland Railway 0-6-4T No. 2000, probably at Trafford Park, Manchester, c. 1912. *Rail Archive Stephenson*

| August 1904 | O/C | 2-6-4T | Belpaire boiler, Stephenson valve gear 80.6 tons |
|---|---|---|---|
| October 1904 | O/C | 4-6-4T | Belpaire boiler, 86.2 tons |
| June 1906 | O/C | 2-6-4T | H round topped boiler, Walschaerts vg, 77.0 tons |
| | I/C | 2-6-2T | H boiler, 'split' side tanks, 73.1 tons |

Three very similar inside-cylinder 0-6-4T schemes, all utilising the rather unsatisfactory but currently standard H round-topped boiler, having alternative side tank arrangements, were also drafted during June 1906, of which the most conventional option, weighing 72.4 tons, was adopted for construction during the following year. This amounted to little more than a tank engine version of the Johnson Class 3 0-6-0, with slightly larger (5ft 7in) coupled wheels. Indeed, but for the substitution of a leading coupled axle, which was held in Cortazzi slides, in place of the leading bogie, it was not that far removed from the 4-4-4T that Johnson had proposed three years earlier. The compact June 1906 2-6-4T proposal (below) would have been the most promising, and greatly superior to the ponderous 0-6-4Ts that were actually built. Given the weight restrictions that prevailed upon the Midland Railway, with an estimated 86 tons it is very surprising that the 4-6-4T was proposed at all. In its

Midland Railway proposal for outside-cylinder 2-6-4T, June 1906.

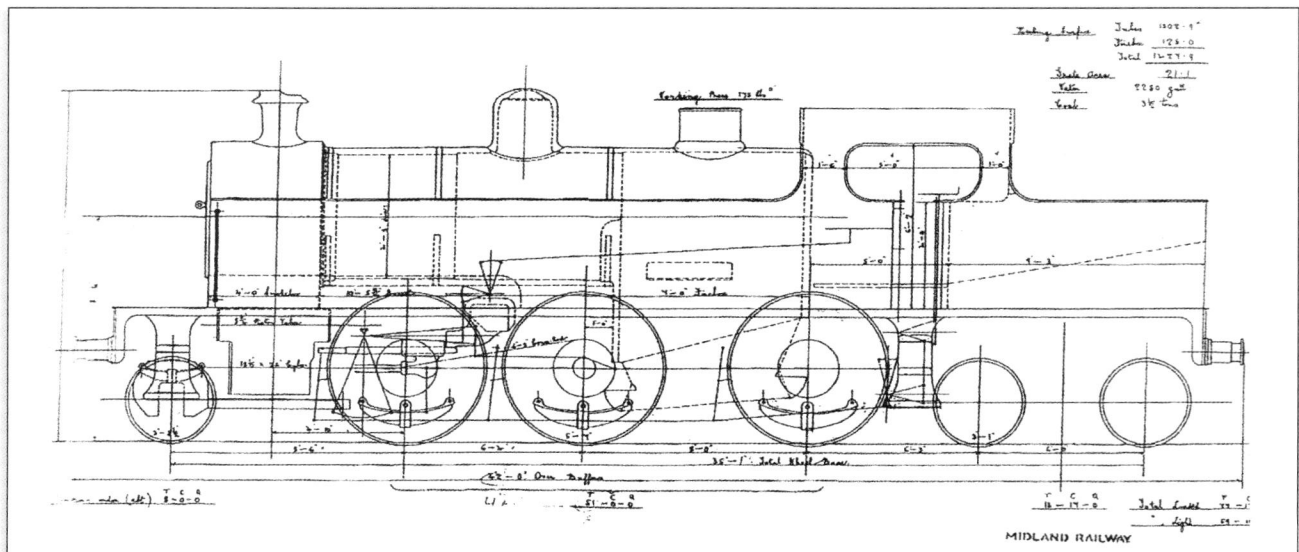

Metropolitan Railway Class G 0-6-4T No. 97 *Brill* at Neasden, carrying later lettering.
*F. H. Stingemore/Rail Archive Stephenson*

general appearance this was remarkably reminiscent of the smaller Class C30 4-6-4T that had been first built in 1903 for the New South Wales Government Railways by Beyer, Peacock & Co. There had also been a proposal for a rigid-frame compound 2-4-4-2T, likewise with 5ft 7in diameter coupled wheels, which would have utilised the Deeley Compound/990 class 4-4-0 boiler. A very similar engine, but with 5ft diameter wheels was also proposed for goods working.

When new, the 0-6-4Ts were particularly employed around Manchester and Birmingham, rather than in the London area. In June 1910, after Deeley's retirement, a superheated 2-6-4T development of these was proposed, on which the already extremely prominent side tanks would have been further extended slightly further ahead of the smokebox. Although this was not proceeded with, the existing 0-6-4Ts were later rebuilt with standard superheated Belpaire boilers. In LMSR days these suffered from two derailments with fatal consequences, one at Newark in 1928, and the other at Ashton under Hill in 1935, of which the last directly led to their fairly rapid withdrawal from service thereafter.

In late 1913, Harry Wainwright's swansong on the SECR consisted of five moderately proportioned superheated 0-6-4Ts (Class J), after various earlier proposals for larger superheated 4-6-2 tank engines had been rejected by the civil engineer.

Although the Metropolitan Railway eliminated steam working with its classic Beyer, Peacock 4-4-0Ts in inner London in 1905, it continued to order new and much larger tank locomotives for a further twenty years to power its main line passenger and freight operations as far north as Aylesbury, which served its fast-developing real estate in 'Metroland'. In September 1914 it called on the private locomotive builders to design and build a tank locomotive capable of working passenger trains of 250 tons and goods trains of 650 tons. It is known that NBL submitted two alternative schemes for 0-8-4Ts, and Beyer, Peacock & Co. a four-cylinder 4-6-4T weighing 95 tons, that was clearly based on a recent delivery it had made to Holland. However, these submissions were each too heavy and the small Yorkshire Engine Company won the day with an altogether lighter and more compact inside-cylinder 0-6-4T (Class G). Two of these were ordered in September 1914 and delivered just over one year later, which were quickly followed by two more engines in early 1916. Of imposing appearance, they were not an unqualified success, proving to be particularly prone to badly cracked frames, which very soon resulted in proposals for outside-cylinder 4-4-4Ts for purely passenger duties. Utilising the same boiler, eight 4-4-4Ts were delivered by Kerr, Stuart & Co. during 1920–21.[4]

The first British tank locomotives to combine six coupled wheels with a *leading* bogie were ten 4-6-0Ts built by the North Eastern Railway at Gateshead Works during 1907–08, specifically to work passenger trains over the heavily graded coastal line between Scarborough and Whitby. They were authorised in January 1907 when their estimated weight had been calculated to be 60½ tons, although when built this was now given out as 69 tons, although unusually the NER diagrams omitted the individual axle weights. However, an undated document

---

[4] Goudie, F., *Metropolitan Steam Locomotives*, Capital Transport, 1990, pp. 42–7.

North Staffordshire Railway Class F 0-6-4T No. 114. This was a larger-wheeled version (5ft 6in) introduced in 1916, of the NSR Class New C 0-6-4T (5ft) introduced in 1914. Its remarkably similar appearance to the Metropolitan Class G (on previous page), which was also painted deep red, would appear to have been entirely coincidental. *Rail Archive Stephenson*

North Eastern Railway Class W 4-6-0T (1907). *Former Ian Allan Archive*

London Brighton & South Coast Railway Class J superheated 4-6-2T express passenger locomotive No. 325 *Abergavenny* (1910). *Former Ian Allan Archive*

came to light some years ago that indicated these but without providing the *total* weight, which was 74½ tons. Furthermore, the axle load on the trailing coupled axle was shown as being no less than 20 tons. This could account for the temporary deployment of the class inland in the Leeds and Harrogate areas when newly built, while the necessary bridge strengthening, etc, was being carried out near the coast. The 4-6-0Ts were later rebuilt as 4-6-2Ts between 1914 and 1917, ostensibly in order to increase the formerly inadequate coal bunker capacity from 2½ to 4 tons. The published total weight in this form (78 tons) was still somewhat suspect, with the maximum axle load now being given as 17 tons. Other than the works official portraits, photographs of these locomotives when first running as 4-6-0Ts are extremely rare, and they were in fact the only British tank engines ever built with this wheel arrangement.

The 4-6-2 passenger tank engine made a simultaneous first appearance in December 1910 in two very different forms, superheated with outside cylinders and non-superheated with inside ones, on the LBSCR and LNWR respectively. Although six units had been authorised, the Brighton engine was built as a singleton (Class J) and was a significant advance in power over the excellent Marsh I3 4-4-2Ts. A second engine was also very soon put in hand, although its completion was delayed by Marsh's successor, Lawson Billinton, pending modifications. Most prominent was the substitution of outside Walschaerts valve gear for internal Stephenson; this locomotive finally

emerged from Brighton Works in March 1912, when it made a trial trip to Littlehampton. The two engines, numbered 325 and 326, enjoyed long lives, throughout which they each retained their original boilers, until they were both retired together in June 1951.

Thirty-seven 4-6-2Ts were ordered from Crewe Works during 1910, the first twenty in February, when superheating was still experimental and before even the first superheated 4-4-0s were ordered. Although the first two 4-6-2Ts were not originally superheated, the next ten were, although ten more non-superheated engines then followed, which were later converted. A final batch of nine was built in 1916. The 4-6-2Ts broke new ground on the LNWR in that, except for two Webb Jubilee compound 4-4-0s already experimentally so fitted, they were fitted with Belpaire boilers.

The Great Central quickly followed suit, in late July 1910 a 4-6-4T, presumably not superheated, was outlined, which was quickly followed a few days later by an alternative superheated inside-cylinder 4-6-2T. Ten of the latter were ordered almost immediately, being intended specifically to handle the increasing suburban passenger traffic operating out of the GCR's Marylebone, London terminus and into the Chilterns.

The LBSCR, LNWR and GCR had each progressed in only a very few years from 4-4-2T to 4-6-2T, but the small LTSR, after its very considerable head start with the 4-4-2T back in 1880, then went one better. When Thomas Whitelegg retired he was immediately succeeded on 1 August 1910 as

LNWR 4-6-2T No. 1021 at Manchester, London Road Station, c. 1920. *P. F. Cooke/Rail Archive Stephenson*

GCR Class 9N 4-6-2T No. 450 climbs away from Rickmansworth on a down outer suburban passenger train, c. 1920. *A. L. P. Reavil/Rail Archive Stephenson*

locomotive superintendent at Plaistow by his son Robert. Only two days later, on 3 August, a preliminary diagram for a 4-6-4T was logged in the Plaistow drawing register. Kenneth Leech was later shown this by Robert Whitelegg during the course of his interview for his unique premium apprenticeship under him at Plaistow Works. When compared with the latest 4-4-2Ts, the 4-6-4T's increased their adhesive weight via three coupled axles. Reduced coupled wheel diameter from 6ft 6in to 5ft 9in, and an increase in boiler working pressure from 170 to 200lb, would all make for greatly enhanced acceleration characteristics and permit the haulage of heavier trains. Traffic had steadily increased on the Tilbury line over the previous thirty years as is indicated below, having roughly doubled between 1900 and 1910 alone. Just how the LTSR managed rather to handle this massive increase in traffic during this latter period, with only a further ten new 4-4-2Ts added to the already existing stock of sixty, is not easy to explain.

### Table 24 The growth of passenger traffic on the LTSR, 1880–1910

| Year | Passenger train mileage | Total number of passengers |
|------|------------------------|---------------------------|
| 1880 | 358,831 | 2,562,231 |
| 1890 | 739,476 | 6,889,773 |
| 1900 | 1,150,071 | 15,774,822 |
| 1910 | 2,057,713 | 32,222,943 |

Taken from *Board of Trade Annual Railway Returns*.

LTSR 4-6-4T, as Midland Railway No. 2101, with replacement Derby chimney. The original was heavily flared with a polished brass cap. *Rail Archive Stephenson*

Eight 4-6-4Ts were eventually ordered from Beyer, Peacock & Co. in October 1911, with the option of using either saturated steam (at £3,800 per engine) or superheated (£4,092), and also in the very real knowledge that the LTSR might soon be taken over, either by the Midland or Great Eastern railways. Since 1910 some major changes had been made to the original scheme: coupled wheel diameter was now to be 6ft 3in, and a Schmidt superheater was decided upon.[5] These locomotives were all completed just over one year later and despatched from Manchester in pairs during December 1912, a few months after the LTSR had indeed been absorbed by the Midland Railway in the previous August. They were therefore finished in crimson lake passenger livery and numbered 2100–7 without bearing names.[6]

Although Robert Whitelegg's official Christmas card for 1912 reproduced the maker's newly taken official portrait of No. 2101, the 4-6-4Ts did not undergo their acceptance trials in Essex until April–May 1913 with Robert Whitelegg himself at the regulator, despite the fact that he had already officially retired on 31 March.[7] Kenneth Leech accompanied him on the footplate for the trials with No. 2102 on 16 April. Always a stickler for accuracy, Leech later was adamant that a speed of 94mph had been touched between Barking and Upminster 'to

[5] For some reason the first engine had its superheater removed in 1914, only for it to be reinstated in 1922.

[6] A fine 1/12 scale model of a 4-6-4T in its would be LTS green livery, and made at Plaistow Works c. 1912, is held by the National Railway Museum at York.

[7] *The Railway Gazette*, 11 April 1913, p. 467, retirement of Robert Whitelegg.

the tune of violent lurching'.[8] The precise whereabouts of these eight 4-6-4Ts between January and March 1913, i.e. after they had left the builders in Manchester, yet before their trials took place, remains a mystery.[9] Presumably they must have been stored at Plaistow.

Such large locomotives were anathema to Derby. Not only that, but their official weight of 94 tons prompted the Great Eastern Railway to forbid their access over its metals into Fenchurch Street Station. It seems possible that the GER had only gained advance notice of the 4-6-4Ts via a small diagram that was published in *The Locomotive* for August 1912, although this had given no (anticipated) weight details. Whitelegg himself later admitted that at an early stage he had entertained concerns on this score, but had been told by his general manager, a former civil engineer, to go ahead.[10] During construction steps were apparently taken to reduce weight via thinner boiler plates and reduced boiler pressure. The 4-6-4Ts in effect immediately became white elephants, as they could not be usefully employed as had originally been intended. The Midland allegedly attempted to sell them on; there is indeed fragmentary contemporary published evidence that the final engine was given trials by the GWR in the Bristol area, and by the SECR in late 1913.[11] During the First World War they usefully hauled heavy coal trains on the Midland main line between Wellingborough and Brent, sometimes incongruously piloted by a Midland 4-2-2.

Robert Whitelegg's subsequent employment during the war was somewhat varied, but at the very end of it he was appointed locomotive superintendent of the Glasgow & South Western Railway, following the decease while still in office of Peter Drummond. Undaunted by the distinctly mixed fortunes of his first 4-6-4Ts in England, he proceeded to build six even more flamboyant 4-6-4Ts for the GSWR in 1922. Whitelegg later claimed that these had prompted George Hughes to produce a 4-6-4 tank version of his new superheated four-cylinder 4-6-0s.[12] Somewhat improbably, it has even been quite validly suggested that the original Tilbury engines had prompted the Furness Railway to invest in its five 4-6-4Ts that were built in 1920, albeit with inside cylinders and without superheaters.[13] It could truly be said that quite coincidentally the Tilbury and Furness 4-6-4Ts could both trace their origins, via much smaller 4-4-2Ts, directly back to William Adams.

British 4-6-4Ts were always a rarity, and none of the thirty-six built in total lasted for very long, at least as tank engines. The only others to be built before 1915 were two magnificent examples by the LBSCR (Class L), which interestingly appear to have been first proposed in December 1912, just as the Tilbury Baltics were being

[8] Leech, K. H., *Loco Profile 27, Tilbury Tanks*, Profile Publications, 1972.

[9] Atkins, P., Baltic Mysteries, *Back Track*, August 2013, pp. 488–93.

[10] Whitelegg, R. H., *Journal of the Stephenson Locomotive Society*, October 1951, p. 270.

[11] *The Locomotive*, January 1914, p. 15.

[12] Whitelegg, R. H., Glasgow & South Western Railway Notes, *Journal of the Stephenson Locomotive Society*, October 1951, pp. 267–71.

[13] Yeomans, G., The Furness Baltic Tanks, *Cumbrian Railways* (Cumbrian Railways Association), October 1997, pp. 59–62.

LBSCR Class L 4-6-4T No. 327 *Charles C. Macrae* at Redhill in 1923. *O. J. Morris/Rail Archive Stephenson*

delivered. The first, No. 327, was completed at Brighton Works in April 1914, to be followed by No. 328 five months later. It has been suggested that Lawson Billinton had intended to build the latter as a 4-6-0, and that the frames were actually cut accordingly.[14] However, there are no such entries relating to a corresponding 4-6-0 tender engine to be found in the Brighton drawing register, which is now held in the NRM archives at York.

Very soon after its first entry into traffic, No. 327 suffered from a series of derailments, such that it, together with the newly completed No. 328, were both put into storage in late 1914 for several months. This particularly followed the derailment of No. 327 at Eastbourne on 27 November, only a fortnight after the engine had experimentally worked north of Willesden Junction and over the LNWR as far north as Rugby, the LNWR being known as 'The Premier Line' on account of the high quality of its permanent way. The following year several alterations were made to the 4-6-4Ts, including the resort to well tanks between the frames in order to lower the centre of gravity, while still retaining intact for appearance's sake the now only partially utilised side tanks. Five more 4-6-4Ts were ordered very soon after the war had ended, and these were completed at Brighton Works between October 1921 and April 1922. An unusual feature of the LBSCR 4-6-4Ts, following on from the second 4-6-2T No. 326, was that the internal piston valves were worked by external Walschaerts valve gear via rocking levers.

Harold Holcroft later recorded that, the very day *before* the Sevenoaks derailment in August 1927 involving a Maunsell 2-6-4T, he had got out a scheme to rebuild the Brighton Baltics as 4-6-0 tender locomotives, which were also to be fitted with Urie N15 boilers.[15] A few years later, when made redundant by the progressive third rail electrification on the Southern Railway, they were indeed rebuilt to become 4-6-0 tender engines between 1934 and 1936, although they still retained their original boilers. As a result, the mythical Billinton 4-6-0 materialised after all, and as such the seven engines would survive for another twenty years until the mid-1950s.

In Britain both 4-6-2Ts and 4-6-4Ts were built in relatively small numbers over a limited period, which ended for both in the mid-1920s. On the other hand, 2-6-2 and 2-6-4 tank engines were built in much greater numbers over a far longer period, with the final examples of both types being turned out by British Railways in 1957. The first 2-6-2Ts had been built seventy years earlier for the Mersey Railway in 1887, upon which they were rendered redundant by electrification in early 1903. Oddly enough, this was shortly before the type appeared a few months later in very divergent forms, and with very different fortunes, on two major main line railways.

The considerable success of the 2-4-2Ts on the LYR prompted John Aspinall's successor, Henry Hoy, to go one better and build a 2-6-2T, with particular regard to working on his company's heavily graded suburban lines in north Manchester to Oldham, etc, which included the formidable 1 in 27 Werneth incline. Still retaining inside cylinders, twenty of these 2-6-2Ts were built at Horwich Works during 1903–04, having boilers that were closely derived from those of the Aspinall 4-4-2 and 0-8-0 tender locomotives. They were not successful, and were soon superseded by further batches of 2-4-2Ts. They were then relegated to shunting duties for which their 5ft 8in coupled wheels were utterly unsuitable, while their long-coupled wheelbase rendered them prone to frequent derailments. Two were withdrawn from service by the LYR as early as 1920, probably on account of having sustained cracked frames, despite one of them having received a new boiler only two years earlier. One of the least known British locomotive classes, the remainder were swiftly withdrawn by the LMSR between 1923 and 1926.[16]

The pioneer GWR 2-6-2T No. 99, completed at Swindon barely one month before its LYR counterpart No. 387, could scarcely have been more different, not only in terms of its design, but also in terms of the total number of successors ultimately built and its sheer longevity. Although in later years regarded as the archetypal GWR suburban passenger tank locomotive, it would appear that this role had originally been envisaged in 1901 for the, in the event, relatively short-lived 4-4-2T, and the 2-6-2T was initially actually proposed as a goods/mixed traffic engine. An early GWR posed official photograph shows one running bunker first on a goods train. No. 99 was tested extensively before series production began in 1905, which resulted in some modifications to the side tanks, while the design was quickly uprated in 1906 when further engines were now fitted with the larger No. 4 in place of the No. 2 Standard taper boiler. A total of ninety-one large 2-6-2Ts was built at Swindon until 1908, yet after a twenty-one-year lapse, construction was resumed under Charles Collett in 1929, and even continued under his successor Fred Hawksworth until 1949!

Much scaled down 2-6-2Ts of the 44XX and 45XX classes, having similar design characteristics, which were intended for light branch line duties, were introduced in 1904 and 1906. While only eleven of the former, with 4ft 1½in coupled wheels, were built, no fewer than 175 45XX/4575 engines with 4ft 7½in wheels, were built up to 1929.

The large 4-6-4Ts ordered by the LTSR and referred to earlier were an embarrassment to the Midland, which nevertheless recognised that six-coupled and more powerful locomotives than the latest 4-4-2Ts were required for the Tilbury Section, pending the MR's (unfulfilled) promise to

[14] Bradley, D. L., *The Locomotives of the London Brighton & South Coast Railway*, Part 3, The Railway Correspondence & Travel Society, 1974, p. 143.

[15] Holcroft, H., *Locomotive Adventure*, Ian Allan Ltd, 1962, p.143.

[16] Mason, E., *The Lancashire & Yorkshire Railway in the Twentieth Century*, Ian Allan Ltd, 1954, pp. 130–3.

Lancashire & Yorkshire Railway 2-6-2T No. 1450. This was the last of the twenty engines to be built, in August 1904, and featured among the last five to be withdrawn from service during August 1926. *Former Ian Allan Archive*

The pioneer Great Western Railway Large 2-6-2T No. 99 as newly built in 1903. Remarkably, direct descendants of this locomotive were still being built at Swindon Works no less than forty-six years later, in 1949. *Former Ian Allan Archive*

electrify it. In February 1914, J. E. 'Jock' Henderson at Derby schemed a more compact and lighter (82-ton) 2-6-4T with 6ft 3in diameter coupled wheels (as on the 4-6-4Ts), incorporating SDJR 2-8-0 pattern 21in by 28in cylinders together with similar short-travel valve gear, and utilising an existing Derby standard superheated 4-4-0 boiler. Only a few weeks later, on leaving Derby to return to the SECR at Ashford, James Clayton took a copy of this diagram with

him, which was very soon reworked there by William Hooley with smaller diameter (19in) cylinders, *long*-travel Walschaerts valve gear, 6ft diameter coupled wheels, and a taper boiler. The boiler was of *Derby* and not Swindon origin, being closely based on that outlined in 1911 for both alternative proposals for a 2-10-0T and 0-6-6-0T for banking on the Lickey incline.[17] This 2-6-4T became the Maunsell SECR Class K, from which the more familiar Class N mixed

[17] Cook, A. F. *Raising Steam on the LMS, the evolution of LMS locomotive boilers*, The Railway Correspondence & Travel Society, 1999, pp. 40–1.

South Eastern & Chatham Railway Class K 2-6-4T No. 790, completed at Ashford Works in 1917. At that time it could claim to be the most modern steam locomotive design in the country. *Former Ian Allan Archive*

traffic 2-6-0 was directly derived, although as mentioned earlier the respective prototypes could not be completed until 1917.

Even then, *series* production of the 2-6-4T only began in 1924 by the Southern Railway, with the engines being named after rivers in southern England, only for these to be dogged by bad riding problems that sometimes resulted in derailments, partly as a consequence of the condition of the permanent way. This culminated in the fatal accident at Sevenoaks in August 1927. This promptly terminated the use of these engines as tank engines, which were then swiftly rebuilt as nameless 2-6-0 tender engines. During that same year the LMSR introduced a new design of 2-6-4 passenger tank engine, which arguably could also trace its origins back to the Derby 1914 proposal. From this the later standard Stanier, Fairburn LMSR and Riddles BR standard 2-6-4Ts were also in effect descended, and together these eventually totalled precisely 800 engines. The last example built at Brighton in March 1957, BR No. 80154, remarkably had an unmistakeable close affinity with SECR No. 790 built at Ashford almost exactly forty years earlier.

Although they functioned as such for less than four years, the only British *eight*-coupled passenger tank locomotives were the first eleven Great Northern 0-8-2Ts built during 1903–04 for service on the joint GNR/Metropolitan widened lines. The initial engine, No. 116, was simply a straight tank engine version of the recent Ivatt 0-8-0 goods engines with a 4ft 8in diameter boiler, but this ensemble was almost immediately deemed to be too heavy at 79 tons. After only two months the engine was accordingly rebuilt with a much smaller boiler and its extremely long side 2,000 gallon tanks were reduced in size. Weight was now only 70¼ tons, and water capacity 1,500 gallons, although the boiler still appeared *superficially* to be of the original size. Ten more engines to this later pattern followed in late 1904, likewise fitted with short chimneys and condensing gear. They were notorious for their extremely rapid acceleration from rest, but in 1907 they were all sent north and had their now superfluous condensing rear removed, before joining the another thirty 0-8-2Ts that had been built in the meantime for goods working, and which were based in West Yorkshire and Nottinghamshire.

Great Northern Railway 0-8-2T No. 116 as originally built in June 1903 with large (0-8-0) boiler, and long side tanks. In this short-lived form the engine weighed only 1 ton less than the 0-10-0WT 'Decapod' that had been built by the GER the previous year. *Former Ian Allan Archive*

After losing weight, GNR 0-8-2T No. 116 is seen after rebuilding only two months later in August 1903 with smaller (4-2-2) boiler, and shortened side tanks. The change of boiler was not readily apparent externally except for the differing position of the dome. *Former Ian Allan Archive*

# Goods tank engines

The archetypal goods tank engine was the 0-6-2T, which on several railways had a direct 0-6-0 tender counterpart, for example on the LNWR and NER. It was particularly popular in South Wales, for whose various local railway companies many were built between 1885 and 1921 mainly to work coal trains down the valleys, 60 per cent of it for export.

The pioneer was the dominant Taff Vale Railway, for whom 208 0-6-2Ts were built throughout this period, concluding with the distinctly bullish Class A, of which the pilot batch was built by R & W Hawthorn Leslie & Co. in

1914. A total of fifty-eight of these were constructed by four different British builders up to 1921, although six of these had originally been ordered from Hanomag in Germany in 1914, only to be cancelled following the declaration of war and the order was placed instead with NBL. Other 0-6-2Ts were also built after 1900 for the smaller Rhymney, Barry, Brecon & Merthyr, Neath & Brecon and Rhondda & Swansea Bay railways.

The North Staffordshire Railway was also a particular devotee of the 0-6-2T, many of which were built in its own Stoke Works between 1908 and 1923. The celebrated

North Staffordshire Railway Class New L 0-6-2T No. 1 leaves Derby for Stoke with a train formed of former LNW stock. This engine was actually built at Stoke Works in early 1923, before the NSR was belatedly absorbed into the LMSR in July 1923. The twenty-eight engines of this class constituted the largest single group of locomotives on this very enterprising small railway. *Rail Archive Stephenson*

Class L, although originally intended as a goods engine, with 5ft diameter coupled wheels in practice proved to be a very useful general purpose locomotive. Several of these were later sold out of service by the LMSR in the 1930s to collieries in the Manchester area, where some remained active until well into the 1960s, although the last former NSR engines to remain in *main line* service, a pair of 0-4-4Ts, had been retired back in 1939.

Of the larger railways, the LBSCR built 134 0-6-2Ts at Brighton Works between 1891 and 1905, many of which lasted until the early 1960s. The North British Railway obtained its first 0-6-2Ts in 1909 and ultimately possessed seventy-five, in addition to which a final thirty were delivered 'posthumously' in 1923.

By comparison, the goods 0-6-4T was much rarer. The short-lived Lancashire Derbyshire & East Coast Railway obtained six of these from Kitson & Co. in 1904. These worked its export coal trains from the Nottinghamshire & Derbyshire coalfield over the Great Central Railway beyond Lincoln to Grimsby. They were developments, with much increased coal and water capacity, of the earlier (1895) Class A 0-6-2Ts from the same builder, having very similar cylinders and Allan straight link valve motion. They were also originally fitted with curious cylindrical extensions to the smokebox in order to house spark arresting equipment. Three more 0-6-4Ts were later ordered, which were delivered in the final days of 1906 just as the LDECR was about to be absorbed by the Great Central. However, these were provided with conventional, slightly extended built up smokeboxes, water pick up, and were *already* fitted with elegant Robinson-style GC chimneys in anticipation of the takeover, which would also very soon be fitted to the 1904 engines.

These 0-6-4Ts do not appear to have remained on the Grimsby workings for very long, being superseded by Robinson GC 0-8-0s and later 2-8-0s. However, a tank engine, given adequate fuel and water capacity, could return with empties from Immingham more quickly without needing to be turned there. In August 1913 a small-wheeled superheated 4-6-2T was first mooted at Gorton for the task, but this was followed two months later by an alternative proposal for a bulky inside-cylinder 2-6-4T. Its unusually high 3,000-gallon water capacity, and large 4½-ton capacity coal bunker supported by a massive 7ft 6in wheelbase trailing bogie, were all very reminiscent of the earlier 0-6-4Ts. It is reasonable to suppose that Robert Thom, the final locomotive superintendent of the former LDECR, who had been at least nominally responsible for the latter, and who since 1907 had been the assistant works manager at Gorton Works, had a hand in this. Twenty 2-6-4Ts were ordered, of which the first two entered traffic during December 1914, by which time the export coal traffic had collapsed on account of the war. Nevertheless, the remaining eighteen engines continued to be built, only gradually emerging between April 1915 and May 1917, but in the event the class was never employed on its originally intended duties, even after the resumption of coal shipments post-1918. The GCR 2-6-4Ts, popularly dubbed 'Crabs', were never regarded as entirely satisfactory, not least on account of their inadequate braking capacity, despite their trailing bogies also being braked when they were first built.

Two rather similarly proportioned goods 0-6-4Ts made their appearance in mid-1914, on the Barry and North Staffordshire railways, both of which were painted red. Non-superheated, the Barry Class L was built by R & W

Lancashire Derbyshire & East Coast Railway Class D 0-6-4T No. 29 (1904). This maker's official portrait shows the original but short-lived extended smokebox, which contained a spark arrester, and Kitson/GNR-style chimney. Both of these features quickly disappeared after the Great Central takeover in 1907. Only a single photograph of one of these engines (No. 33) in LDEC black working livery is known to exist. *Former Ian Allan Archive*

Great Central Railway Class 1B 2-6-4T No. 272 (1914). This was intended to undertake similar duties to those of the inherited LDEC 0-6-4Ts (above). With an adhesive weight of 60 tons (with full side tanks), sanding facilities were noticeably more generous than on the earlier 0-6-4Ts. *Former Ian Allan Archive*

Hawthorn Leslie, and numbered ten engines, which quickly showed a propensity to derail at certain locations. This later prompted unrealised proposals for a superheated outside-cylinder 2-6-2T, somewhat on Swindon lines. Although seven of the 0-6-4Ts were rebuilt with new boilers during 1922–24, four of them GWR standard taper boilers, possibly owing to a misunderstanding all ten engines were later condemned by the GWR almost at a stroke as early as 1926, after very short lives.

The superheated NSR New C 0-6-4T had an otherwise outwardly very similar Belpaire boiler to the Barry engine, and likewise featured a cylindrical smokebox resting on a saddle. It was a logical enlargements of the existing Class L 0-6-2Ts, which remained in production at Stoke until 1923. Undoubtedly delayed by the outbreak of war, a larger-wheeled (5ft 6in) of the 0-6-4T, for passenger duties, Class F, first appeared in 1916. No other British railway ventured beyond a single design of 0-6-4 tank.

In 1910–11 the NER built twenty powerful three-cylinder 4-6-2Ts (Class Y) for short haul work in the Durham coalfield, from the mines to the coal staithes on the coast. This was very much a feature of its operations, as much Durham coal was shipped to London by sea for gas-making purposes.

The prototype GWR 2-8-0T No. 4201 (1910). Although this class was superheated from the start, top feed did not begin to make its appearance on GWR locomotives until the following year. *Former Ian Allan Archive*

*Eight*-coupled goods tank engines were always distinctly uncommon other than on the GWR. In 1903 the Caledonian Railway designed and built six inside-cylinder 0-8-0Ts, mainly for branch line duties, the 492 class. Two years later, George Churchward is reputed to have considered a 2-8-2 tank version of his new 2-8-0 for service in South Wales. In 1907 there were proposals for a more compact 2-8-0T that would carry either the No. 2 or slightly larger No. 4 standard taper boiler. The latter was not progressed until 1910, by which time the engine was

to be superheated, although top feed had not yet quite arrived. Ultimately Swindon built no fewer than 205 2-8-0Ts up to 1940, although some more when initially completed as such in 1929 were stored and later extended in 1934 into 2-8-2Ts with 6-ton coal bunker capacity.

In 1907, Richard Deeley on the Midland Railway proposed a rather stately four-cylinder compound 2-8-2 goods tank with full-length side tanks, which was reconsidered a few years later for banking purposes at Bromsgrove (see Chapter 13).

Midland Railway Deeley 0-4-0T No. 1528, built at Derby Works in 1907. *Former Ian Allan Archive*

Great Eastern Railway Hill 0-4-0T No. 227, built at Stratford Works in 1913. *Former Ian Allan Archive*

## Shunting and banking tanks

Short-wheelbase 0-4-0 and 0-6-0 tank engines, with outside cylinders and often fitted with saddle rather than side tanks, featured on railways that served dock facilities. That said, in 1907, the Midland Railway, which had previously built a number of diminutive inside-cylinder 0-4-0STs, introduced a larger new 0-4-0 side tank with outside cylinders and external Walschaerts valve gear. A similar format was followed by the GER in 1913. The heaviest *railway*-owned 0-4-0Ts were six little-known engines of the 272 class that were built by the Glasgow & South Western Railway at Kilmarnock between 1907 and 1909, which weighed in at 40½ tons. In 1902 the GWR built a solitary oil-burning 0-4-0T with Lentz boiler that was numbered in the prototype range as 101. Although actually employed as a works shunter at Swindon, the original intention had been to use it in ultra-light branch line passenger service. It was converted to coal burning with a conventional boiler in 1905 but was scrapped in 1911.

In 1910 the Caledonian Railway produced a new design of short wheelbase 0-6-0 side tank with outside cylinders, which had originally been proposed as a saddle tank. Almost simultaneously the GWR built five small 0-6-0STs to a design first built by Sharp, Stewart & Co. for the Cornwall Minerals Railway in 1874. These were the only 0-6-0Ts to be built for the GWR between 1906 and 1928. For trip working and shunting the GNR and LBSCR each introduced new designs of 0-6-0 side tank in 1913; a superheater was even fitted to one of the GNR engines when new, but this was found not to be justified.

The only Scottish 0-6-4Ts were eight built by NBL for the Highland Railway between 1909 and 1911 specifically for banking purposes on its heavily graded main line between Perth and Inverness. (Large tank locomotives were extremely rare in Scotland, although latterly the Caledonian Railway invested in twelve 4-6-2 passenger tank engines in 1917, and the GSWR in six 4-6-4Ts in 1922.)

For their part, eight-coupled locomotives were even more uncommon in southern England, but in 1905 the LBSCR proposed to turn out the last two of its twelve E6 0-6-2Ts as 0-8-0Ts instead, for which even a general arrangement drawing was made. In the event, Nos 417/8 appeared as 0-6-2Ts after all, but were nevertheless characterised by the heavier section coupling rods that had been specially designed and already fabricated for them.

As mentioned earlier, the GNR built forty-one 0-8-2Ts during 1903–06, derived from its 0-8-0s, which spent most of their lives on coal workings in West Yorkshire and Nottinghamshire. From 1913 these ungainly engines were all rebuilt with the larger 0-8-0 boiler originally intended, of which a select few were also superheated.

In March–April 1908 the LYR built five 0-8-2T heavy shunting and banking tanks. Based on the Aspinall 0-8-0 goods engines, their boilers were of considerably increased diameter, and set the general pattern for those that were later fitted to the Hughes 0-8-0s. The cylinders were 21½in diameter, giving a starting tractive effort of 34,000lb, which was on a par with that of the contemporary three-

Lancashire & Yorkshire Railway 0-8-2T No. 1505, built at Horwich Works in April 1908. *Former Ian Allan Archive*

cylinder Great Central 0-8-4 and North Eastern 4-8-0 tanks to be described. One of the class was regularly employed rendering assistance over the 2-mile Baxenden bank near Accrington, which contained gradients of 1 in 38, i.e. as severe as those encountered on the Lickey incline on the Midland Railway near Bromsgrove.

The LNWR followed suit three years later with a 0-8-2T similarly based on its own Class G 0-8-0 tender locomotives,

and thirty of these were built between late 1911 and 1917. No fewer than forty more were ordered in February 1921, but later cancelled. In their place thirty superheated 0-8-4 tank versions of the new G2 0-8-0 heavy goods engines were ordered in September 1922, particularly with operations in South Wales in mind. These were turned out of Crewe Works during 1923, some of which when new were briefly tried on Manchester–Buxton passenger trains.

The builder's official portrait of the truly imposing three-cylinder 0-8-4T, which it had both designed and built for the Great Central Railway within six months during 1907. The very necessary steam reverser is just visible.

The slightly smaller North Eastern Railway three-cylinder 4-8-0T, which was possibly prompted by its Great Central counterpart. *Former Ian Allan Archive*

The only previous 0-8-4Ts had been the four three-cylinder locomotives designed and built by Beyer, Peacock & Co. for the Great Central Railway in late 1907 referred to in Chapter 3. It is recorded that at Wath concentration yard, which had been designed to handle 5,000 wagons per day, these engines, stationed at Mexborough, regularly worked a continuous six-and-a-half-day week, from midnight on a Sunday and through to 6am on the following Sunday. They were required to push a maximum of 1,200 tons up an incline of around 1 in 100 at speeds of 9–10 mph. Early LNER locomotive diagrams credited them with a (theoretical) cylinder horsepower of 1,457. However, this was calculated according to an American formula devised by F. J. Cole, which assumed a mean piston speed of 1,000ft per minute, thereby corresponding to 38½mph in this instance, which was hardly relevant to a shunting locomotive. It is a somewhat sobering fact that these giants, which weighed nearly 100 tons, were eventually and rather more quietly replaced in 1953 by modest diesel-electric 0-6-0s of merely half their weight, that were rated at merely 350HP! The four 0-8-4Ts were gradually routinely fitted with standard superheated boilers ex-2-8-0s, although it was generally found in British practice that superheaters were not really effective on shunting locomotives. They were retired between 1954 and 1957.

In December 1907 the North Eastern Railway ordered ten modestly proportioned short-wheelbase 0-6-2Ts of new design for heavy shunting duties, only to cancel them a few months later in favour of a like number of three-cylinder 4-8-0Ts (Class X), which would be the last new locomotives to be built at Gateshead Works. These may possibly have been inspired, at least in part, by the recent three-cylinder

0-8-4Ts on the GCR. George Heppell made no reference to the latter in his memoir, although he did record that he had been responsible for suggesting that all three cylinders should be cast together as a single unit for the 4-8-0T, a task that was entrusted to Kitson & Co. in Leeds. At his request Wilson Worsdell authorised a dynamometer car test, at which Heppell was present, with a new 4-8-0T setting off from Hull for Bridlington, initially with 1,200 tons behind the coal bunker, with which it sustained a maximum speed of 35mph over half a mile of level track.[18]

It is an interesting fact that the first four new-build British three-cylinder simple locomotive designs, beginning with the GER Decapod in 1902 (see Chapter 13), were all tank locomotives. Three cylinders quickly became something of a trademark on the NER, however, being adopted on several new classes during the next few years, both tender and tank, passenger, goods and shunting, including the celebrated Class Z 4-4-2 in 1911.

After 1922 the LNER built five more of the 4-8-0Ts at Darlington in 1925 for 'local' use, and two more 0-8-4Ts at Gorton in 1932 for hump shunting at its new marshalling yard at Whitemoor in Cambridgeshire. The latter were embellished with auxiliary boosters, American-style, on the trailing bogie, following trials on one of the earlier engines. These were later removed from all three engines during 1943.

The production of these two large GCR and NER tank locomotive designs in particular reflected the immensity of the British coal industry pre-1914, and the resultant very heavy rail traffic that it generated in northern England during this period.

[18] Heppell, G., *North Eastern Locomotives; a draughtsman's life*, North Eastern Railway Association, 2012, p. 18.

The two Leek & Manifold Light Railway 2-6-4Ts parked bunker to bunker, with No. 1 *E. R. Calthrop* nearest the camera, at Hulme End in August 1933, a few months before the permanent closure of the line in March 1934 by the LMSR. *Colling Turner/Rail Archive Stephenson*

## Narrow-gauge locomotives

At the other extreme, the British locomotive industry produced a very small number of new locomotive designs for narrow-gauge railways in England, Wales and Scotland during the Edwardian period:

| 1902 | Vale of Rheidol (1ft 11½in gauge) 2-6-2T (2), Davies & Metcalfe Ltd |
| --- | --- |
| 1902 | Welshpool & Llanfair Light (2ft 6in) 0-6-0T (2), Beyer, Peacock & Co. |
| 1904 | Leek & Manifold Light (2ft 6in) 2-6-4T (2), Kitson & Co. |
| 1906 | Welsh Highland (1ft 11½in) 2-6-2T (1), Hunslet Engine Co. |
| 1906/7 | Campbeltown & Machrihanish Light (2ft 3in) 0-6-2T (2), Andrew Barclay & Co. |
| 1908 | Welsh Highland (1ft 11½in) 0-6-4T (single Fairlie) (1), Hunslet Engine Co. |
| 1914 | Southwold (3ft 0in) 0-6-2T (1), Manning Wardle Ltd |

The two Vale of Rheidol 2-6-2Ts were the only locomotives to be built by Davies & Metcalfe Ltd of Stockport, whose actual specialism was the manufacture of injectors. The remaining locomotives were all notable at that period for having been fitted with Walschaerts valve gear, while the two Leek & Manifold engines were the first 2-6-4Ts to work in Britain by ten years. The Campbeltown & Machrihanish Railway was the only narrow-gauge passenger carrying railway in Scotland.

# 12
# Some Odious Comparisons

For the climactic year 1913, from their respective new-style annual reports to the Board of Trade it was possible to make direct and detailed statistical comparisons between such extremes as the huge London & North Western Railway, with more than three thousand locomotives, down to the humble Knott End Railway in rural Lancashire, which operated only four small tank engines and which branched off the LNWR main line at Garstang. Locomotive operating statistics, derived from these, are given below for the fourteen largest companies, which have been ranked by their total route mileage, followed by those for six of the leading smaller railway enterprises.

For the larger railways in many cases average engine miles per annum were surprisingly close, typically around 23–25,000 miles, but ranged between the extremes of 19,000 on the LYR to as high as almost 30,000 on the SECR. The LYR carried a very heavy coal traffic, as did the Hull & Barnsley and Taff Vale railways, and the average annual engine mile figures of all three, although much lower than the average, were almost identical. On the LYR

The first Midland Railway Class 2 4-4-0 renewal, No. 483, with Schmidt superheater (1912). Ninety such renewals were carried out between 1912 and 1914, followed by seventy-five more up to 1924. Between 1928 and 1932 a further 138 engines were built entirely new to this basic design by the LMSR. *Former Ian Allan Archive*

## Table 25 A comparison of locomotive running and repair costs on twenty leading railways

| Railway | Average engine miles per loco per annum | Average fuel cost per engine mile p | Average running cost per engine mile* p (A) | Average repair cost per engine mile p (B) | Average total cost per loco per engine mile p (A + B) | Average total annual cost per loco £ | Gross traffic receipts per loco £ |
|---|---|---|---|---|---|---|---|
| GWR | 23,582 | 1.29 | 2.67 | 1.18 | 3.86 | 920 | 4,952 |
| LNWR | 24,747 | 1.52 | 3.04 | 0.94 | 3.99 | 989 | 5,123 |
| NER | 22,979 | 1.40 | 2.94 | 1.47 | 4.42 | 1,015 | 5,596 |
| MR | 23,520 | 1.27 | 2.74 | 1.00 | 3.74 | 667 | 4,640 |
| NBR | 27,951 | 1.39 | 2.66 | 0.74 | 3.40 | 950 | 4,917 |
| GER | 24,005 | 1.58 | 2.79 | 1.23 | 4.02 | 973 | 4,664 |
| CR | 27,964 | 1.61 | 2.74 | 0.74 | 3.48 | 974 | 5,085 |
| LSWR | 26,557 | 1.89 | 3.23 | 1.00 | 4.24 | 1,101 | 5,603 |
| GNR | 23,118 | 1.59 | 3.15 | 1.03 | 4.18 | 966 | 4,408 |
| GCR | 25,369 | 1.50 | 2.53 | 1.03 | 3.57 | 905 | 3,622 |
| SECR | 29,685 | 2.00 | 3.45 | 0.90 | 4.34 | 1,289 | 6,752 |
| LYR | 19,182 | 1.39 | 3.06 | 1.08 | 4.14 | 821 | 4,115 |
| GSWR | 23,431 | 1.51 | 2.59 | 1.10 | 3.70 | 856 | 3,863 |
| LBSCR | 25,398 | 1.79 | 3.08 | 1.10 | 4.19 | 1,065 | 5,797 |
| **Mean** | **24,866** | **1.47** | **2.88** | **1.02** | **3.90** | **964** | **4,938** |
| HR | 25,457 | 1.80 | 2.78 | 0.47 | 3.25 | 828 | 3,872 |
| GNSR | 24,232 | 1.99 | 2.90 | 0.45 | 3.37 | 809 | 4,426 |
| NSR | 24,628 | 1.28 | 2.66 | 0.84 | 3.50 | 848 | 5,633 |
| FR | 23,440 | 1.09 | 2.55 | 0.52 | 3.07 | 689 | 4,337 |
| TVR | 19,298 | 1.70 | 3.14 | 0.91 | 4.05 | 782 | 3,854 |
| HBR | 19,182 | 1.47 | 2.90 | 0.72 | 3.63 | 719 | 3,865 |
| **Mean** | **22,706** | **1.56** | **2.99** | **0.65** | **3.48** | **779** | **4,331** |

\* This total includes crew wages, fuel, lubricant, sand & water costs, etc.

train miles amounted to only 52 per cent of its total engine miles, yet these earned the highest total revenue per mile of any company, closely followed by the NER. A significant proportion of the balance would have been accounted for by non-revenue-earning shunting operations. (See Appendix 1) However, quoted total shunting engine miles, by its very nature, was not a very accurate statistic, having to be calculated on the basis of a purely notional 5 miles per locomotive hour in steam.

Goods traffic generated considerably more revenue per train mile than did passenger traffic. In 1913 the NER reckoned that on this basis passenger trains on average generated 20.5p, general merchandise 55.4p, and mineral 72.3p per mile.[1]

Fuel costs when averaged across the board accounted for just half of locomotive running costs, but were notably higher in the south of England, being remote from any significant coalfields. Thus in 1908 the LBSCR paid 89.9p per ton, which was very closely followed by the LSWR at 88.8p, while for the SECR the figure was a little lower at 79.0p. This compared with 54.6p by the Midland and 53.6p by the Great Northern, both of these companies

having direct access to the coalfields in the Midlands and South Yorkshire. Coal prices were also high in northern Scotland, also on account of the relative remoteness of the Highland and Great North of Scotland railways from their respective sources in the Lanarkshire and Fife coalfields. (By early 1915 the HR was paying 95p per ton.) This would naturally have a direct bearing on overall operating costs, which nationally were the highest on the SECR, and the lowest on the GCR, although paradoxically the latter, and also despite its heavy mineral traffic, was the least productive of the major railways.

Fuel costs in 1913 averaged around 1.5p per engine mile, but individually ranged from between 1.09p on the Furness to 1.80p and 1.99p on the Highland and Great North of Scotland respectively. The low relative cost of coal as supplied to the Furness Railway was remarkable. Later, this might have accounted for the very surprising post-1918 omission of a superheater in Barrow's specification for a large inside-cylinder 4-6-4T, of which five were built for the FR by Kitson & Co. in 1920.

Locomotive repair costs were noticeably lower on the smaller railways. This was particularly true of the Highland

[1] Irving, R. J., *The North Eastern Railway, 1870–1914: an economic history*, Leicester University Press, 1976.

and Great North of Scotland lines, where lower rates of pay in their respective workshops in Inverness and at Inverurie could also have been a significant factor. Of the major companies, repair costs were lowest on the Midland, probably on account of the fundamental simplicity of the great majority of its large locomotive stock, about 50 per cent which at that time was accounted for by 1,500 non-superheated 0-6-0s. On the other hand, although not so readily explicable, repair costs were surprisingly high on the North Eastern Railway.

On average, and by a fair margin, the oldest locomotives were also those of the Midland Railway, of which only 23 per cent had been built after 1898, compared with the national overall average of 36 per cent. The Great Northern and Great Central railways, on the other hand, boasted the most youthful locomotive fleets with just over 50 per cent of their locomotives having been built during the previous fifteen years, even after disregarding their collective forty American-built 2-6-0s that were already in the course of withdrawal. In fact, if one also excludes the locomotives that had recently

## Table 26 Heavy repairs made to loco stock of the fourteen major railways, 1913

| Rly | Loco stock at 31 Dec 1913 | Number of heavy repairs in 1913 | Number of heavy repairs as % of loco stock | Average engine mileage between heavy repairs |
|---|---|---|---|---|
| GWR | 3,070 | 1,258 | 41 | 58,209 |
| LNWR | 3,084 | 1,842 | 60 | 41,437 |
| NER | 1,998 | 749 | 37 | 61,297 |
| MR | 3,019 | 1,324 | 44 | 53,360 |
| NBR | 1,058 | 340 | 32 | 86,975 |
| GER | 1,274 | 677 | 53 | 45,174 |
| CR | 997 | 435 | 44 | 64,093 |
| LSWR | 937 | 366 | 39 | 67,998 |
| GNR | 1,345 | 507 | 38 | 61,328 |
| GCR | 1,352 | 581 | 43 | 59,035 |
| SECR | 719 | 351 | 49 | 60,807 |
| LYR | 1,547 | 673 | 44 | 46,424 |
| GSWR | 521 | 126 | 24 | 96,966 |
| LBSCR | 600 | 249 | 42 | 53,169 |
| **Mean** | **1,537** | **677** | **44** | **55,372** |

In 1913 the fourteen major railways carried out a total of 9,478 heavy repairs on their combined locomotive stock of 21,551 locomotives, or 44 per cent of these. The actual variation between individual companies in this respect is illuminating. For the LNWR the proportion was exceptionally high at 60 per cent. An average is an average. Crewe had built one thousand new locomotives between 1900 and 1911 alone, about one quarter of them Webb compounds, but in 1913 just on two-thirds of its locomotive stock was comprised of Webb's simple small inside-cylinder tender and tank locomotives, which had been designed very much in the straightforward John Ramsbottom mould. The implication is therefore that a significant number of post-Webb locomotives must have made fairly frequent visits to Crewe Works for major attention. At the other extreme, the GSWR recorded the corresponding and remarkably low proportion of only 24 per cent, with considerably more than double the LNW mileage run between heavy repairs. Second best in this respect was the NBR. Given that at this period boiler maintenance dominated locomotive repairs and largely determined their frequency, could this have been attributable to the legendary soft water supplies for which Scotland was notable? Unfortunately, the corresponding statistics for the Caledonian Railway for their part do not appear to bear this suggestion out.

been absorbed from other companies (thirty-seven from the LDECR by the Great Central in 1907, and ninety-four from the LTSR by the Midland in 1912), the average age of the *mainstream* Great Central locomotive stock was merely thirteen years, compared with no less than twenty-six years on the Midland, for which new locomotive construction after 1901 had progressively fallen to considerably lower levels than that on the LNWR and GWR, whose total stock likewise now topped the three thousand mark.

## Table 27 Median build date of locomotives of the fourteen major railway companies at 31 December 1913 (absorbed and imported locomotives omitted)

| | |
|---|---|
| GCR 1900 | GER 1895 |
| GNR 1899 | LSWR 1893 |
| CR 1897 | GWR 1891 |
| SECR 1897 | NER 1891 |
| LNWR 1896 | NBR 1891 |
| LBSCR 1896 | GSWR 1891 |
| LYR 1895 | MR 1887 |

In fact, Derby Works officially turned out only two new engines for the Midland's own use between 1910 and 1916, although from 1912 a substantial number of existing 4-4-0s (ninety between 1912 and 1914), some of them of relatively recent construction, were very heavily renewed. Their tenders apart, these in effect amounted to being new locomotives; a very senior former Derby man once quipped that from the originals 'they only retained the space between the bogie wheels'. As 'renewals' these did not require official sanction and would also qualify for a £10 discount (which in 1912 was roughly the monetary equivalent of £1,000 a century later) on the now standard £50 royalty for the use of the Schmidt superheater, which was levied for an application to a *new* locomotive. The result was the rather uninspired Midland '483' or 'Superheated Class 2' 4-4-0, whose cylinder and valve gear design, with directly driven outside admission piston valves located below the cylinders, harked back to the 1890s. Like the Class 4 0-6-0, this was perpetuated for new construction with only minor modifications after 1922 by the newly formed London Midland & Scottish Railway. Indeed, the LMSR latterly adopted this rather mundane 4-4-0, now redesignated as Class 2P, as a benchmark (=100) as regards relative repair costs for comparison with some of its other locomotive classes. On this scale the ex-LNWR 'Prince of Wales' 4-6-0 came out at 147. Frame performance, i.e. proneness to cracking, to which the 'Princes' were especially vulnerable, had a particularly strong bearing on this index.

As regards the average *total* annual cost of running a locomotive, the three most southerly based railways in England topped the list mainly on account of their higher fuel costs, then followed by the North Eastern because of its high repair costs. Nevertheless, in terms of gross traffic revenue earned per locomotive, the South Eastern & Chatham Railway was head and shoulders above the rest at £6,752, and still remained so in terms of the *net* revenue (£5,463) after the deduction of its high running and repair costs. This apparently laudable high efficiency was in fact partly due to a current acute locomotive shortage that had resulted from the untimely closure in 1911 of Longhedge Works as a second repair facility, which had been demanded by the railway's directors in their wisdom as an economy measure. The situation was not helped by the loss of four elderly locomotives in collisions, etc, during a remarkably short period, but twenty-two new 4-4-0s would be delivered in 1914. In 1913 SECR locomotives ran an average of 22,000 train miles apiece, which was by far the highest such figure achieved on any of the major British railway companies, and which was 33 per cent above the mean. Also during 1913 the SECR investigated the feasibility of adopting extensive electrification, but with the turn of international events matters did not proceed any further before the company's legal demise in 1923.

Quite coincidentally, as regards high locomotive profitability, pre-eminent among the smaller railways was the North Staffordshire Railway, whose senior locomotive personnel in 1913, i.e. the locomotive superintendent, works manager, and chief draughtsman, had each served on the SECR at Ashford about ten years earlier. This resulted in 4-4-0s, 0-6-0s and 0-4-4Ts in particular appearing on the NSR that had a distinct resemblance to their counterparts on the SECR, only they were painted a deep red rather than dark green. At the other end of the scale, the gross receipts per locomotive on the Great Central were remarkably low at only £3,622, while total expenses amounted to an exceptional 25 per cent of this (the mean was 19.5 per cent). This gave *net* receipts of only £2,717 per locomotive, which amounted to just half that realised on the SECR. Interestingly, the *total* gross receipts of the two companies was actually virtually the same at almost £5 million, but the GCR operated nearly twice as many locomotives as did the SECR, but which to their credit returned the lowest fuel and repair costs per engine mile of any of the major railway companies (see Table 25 and Appendix 2). This rather belied the likewise overtly opulent appearance of the GCR's leading passenger locomotives. Indeed it perhaps gives some credence to the story that its directors were politely requested to defray the cost of casting the brass nameplates for the handsome new 4-4-0s that were built in 1913 that would be named after them!

It was remarkable how close were the respective mileages and costs for the locomotives of the North British and Caledonian railways. Many of these did, in fact, enjoy a considerable mechanical affinity through the long-lasting legacy of Dugald Drummond, who had successively served as the locomotive superintendent on these two companies between 1875 and 1890.

The greatest contrast to be found between any two of the fourteen major railways in 1913, was that between the South Eastern & Chatham in South east, and the Lancashire & Yorkshire in North-west England, which shared the common denominator of running approximately 16 million revenue earning miles during that year:

### Table 28 A statistical comparison of the SECR and LYR in 1913

|                  | SECR       | LYR        |
|------------------|-----------|------------|
| Route mileage    | 624       | 532        |
| Locomotive stock | 719       | 1,577      |
| **Receipts**: £  |           |            |
| Passenger        | 3,679,505 | 2,877,576  |
| Goods            | 1,175,486 | 3,610,222  |
| Total            | **4,854,991** | **6,487,798** |
| No. passengers   | 56,313,033 | 69,170,122 |
| Freight tonnage  | 6,394,249 | 26,724,850 |

# 13

# Four Unique Locomotives

Four special locomotives were designed during the Edwardian period that were truly unique. Each broke new ground, not least on account of their sheer size. Between them they incorporating up to eight cylinders, ten coupled wheels and all of them had unusually large boilers. Designing them required considerable ingenuity at the drawing board. The first locomotive is recorded as having only ever operated on five separate occasions during a period of merely six months. The second covered more than half a million miles in its original almost legendary form. The third underwent trials amid great secrecy, without the knowledge of the contemporary technical press, and it also *almost* completely eluded the camera. Its complex technical details would only be publicly revealed some twenty-five years after its demise. The fourth, from the same design office, in contrast, operated in the public eye for thirty-six years and it was widely photographed from while being under construction in 1919 until actively being broken up in 1957.

## The Decapod

As early as January 1896 Francis Webb, when addressing the Crewe Mechanics Institute, boldly declared, somewhat prematurely as it turned out, 'in ten to fifteen years from now trains moved by electricity will run from all large centres of the country at a rate of speed which can hardly be realised, probably 100mph'.[1] By the dawn of the 20th century the power of electricity was actually beginning to have a detectable *adverse* impact on railways. This was in large urban areas through the rapid proliferation of electric street tramways in such cities as London, Manchester, and Newcastle upon Tyne. Such were now engaged in competition with well-established steam-worked railway suburban services, on which passenger usage as a consequence began to decline, and with it the associated revenue, to the railway companies' obvious concern. In the London area there were also proposals to extend the electrically worked underground ('Tube') railways into Great Eastern Railway territory.

The Great Eastern Railway's London terminus at Liverpool Street in 1902 was handling 220,000 passengers a day from eighteen platforms, of whom 90,000 commuted more than 16¼ miles to and from Enfield, with fifteen intermediate stops of twenty-seven-second average duration. These trains were run at five-minute intervals, worked by diminutive 0-6-0 tank engines that took forty minutes overall, giving an overall average speed of 24.4mph, and an average running speed of 29.4mph. In a bid to accelerate the service in order to compete with the local electric tramways, a revolutionary 0-10-0 tank engine was designed at Stratford under James Holden, whose memorable design brief was to accelerate a 300-ton train from rest to 30mph in thirty seconds.[2] It was calculated that this would require the development of an average indicated horsepower in the cylinders of 1,200. Quite apart from loading gauge considerations, three simple cylinders with their cranks mutually phased at 120 degrees were specified in order to achieve the necessary acceleration. All three cylinders were to be in line abreast beneath the short large-diameter smokebox, driving onto the centre coupled axle, to facilitate which the middle connecting rod would be forked in order to permit the second coupled axle to pass through it. Although inspired by an American precedent, James Holden was nevertheless granted a related patent for this.[3] The middle cylinder discharged through the central blastpipe, while the two outside cylinders discharged through separate concentric annular apertures around the central blastpipe.

[1] Reed, B., *Crewe Locomotive Works and its Men*, David & Charles, 1982, p. 236.
[2] Skeat, W. O., The Decapod Locomotive of the Great Eastern Railway, *Trans. Newcomen Society*, 1951/52 & 1952/53, pp. 169–185.
[3] Holden, J., BP 28,946/1902, (forked locomotive connecting rod).

Great Eastern Railway three-cylinder 0-10-0WT No. 20. *Former Ian Allan Archive*

The boiler was the largest so far fitted to a British locomotive, having a wide firebox that afforded a grate area of 42 sq ft, a figure that would not be exceeded for more than twenty years. The engine was completed at Stratford Works in December 1902, at an approximate cost of £5,000 (the equivalent of about £600,000 at 2018 prices) and the first electrically monitored acceleration tests took place on 11 January 1903 over a stretch of track at Chadwell Heath. Four more tests took place, on 8 February, 29 March, 26 April and finally on 28 June 1903, when the engine ran down Brentford bank at 55mph and also at last achieved its design objective.[4] It did not run again, and it was extremely unlikely ever to have entered series production. Its weight, long rigid wheelbase, and the extremely low capacities for both fuel and water within the very limited space available would have been very serious limitations. However, although seemingly a very expensive project, the locomotive had effectively served its purpose in seemingly heading off more proposed electrically operated competition in suburban Essex.

In May 1904 a proposal to convert the locomotive into a two-cylinder 0-8-0 tender engine was approved, but it was almost two years before the reconstruction began in

March 1906. What emerged in early October 1906 only retained the two outside cylinders and four of the five coupled wheelsets from the original 0-10-0WT. This decidedly awkward-looking ensemble was retired in late 1913, and broken up in June 1914. By this time the GER could boast of the most intensive steam-operated suburban passenger service in the world, which ironically was in the main worked by diminutive 0-6-0Ts! This service was further highly refined in 1920, resulting in the colour-coded so-called 'Jazz Trains'.

## The Great Bear

The world's first true 4-6-2 or 'Pacific' locomotives were built in the USA for service on the 3ft 6in gauge in New Zealand, by the Baldwin Locomotive Works in 1901. The preliminary weight diagram for Britain's first 4-6-2 tender locomotive was prepared by the GWR at Swindon only five years later in June 1906, two months after the completion of four-cylinder 4-4-2 No. 40 *North Star*.[5] At this early stage the 4-6-2 was not even to be superheated, but when completed in February 1908 it was fitted with a Field tube superheater (i.e. with concentric single return

[4] Brooks, L., Another view of the Decapod, *Great Eastern Journal* (GER Society), No. 163, July 2015, pp. 4–9.
[5] Rutherford, M., *'Castles' & 'Kings' at Work*, Ian Allan Ltd, 1982, p. 17.

Great Western Railway four-cylinder 4-6-2 No. 111 *The Great Bear* as built in 1908. Top feed was later fitted in January 1913, and a plain cast iron chimney in 1920. *Former Ian Allan Archive*

elements) that would later be replaced in 1913 and 1920 by more conventional patterns. GWR No. 111 had only been preceded a few months earlier by Europe's first 'Pacific', a compound that appeared on the Paris–Orleans Railway in mid-1907, although in Germany the Baden State Railway had originally intended to have a 4-6-2 (also a compound) in service as early as 1905.

The assembly of the GWR 4-6-2 occurred between the completion of two batches of ten four-cylinder 4-6-0 Stars derived from 4-4-2 No. 40, and it was given the inspired name of a prominent stellar constellation, *The Great Bear*. Its boiler was exceptionally large, attaining a maximum diameter of 6ft, although the grate area was actually fractionally less than that of the GER Decapod. Very much a leap in the dark from the design point of view, by later standards the boiler was not well proportioned as it extended 22½ft between tubeplates. In the final British 4-6-2 designs more than forty years later this dimension amounted to only 17ft, but the combustion chambers on the recent Kruger 2-6-0s had proved to be very troublesome. Although having a water capacity of only 3,500 gallons, no greater than that of some existing standard six-wheel tenders, the entirely new special double-bogie tender, which was shown even on the initial diagram, was intended simply to appear more in keeping with the greater size of the locomotive. Despite the latter measuring 71ft 2in over buffers, 8ft longer than a Star 4-6-0, no reference ever appears to have been made regarding the engine's relationship with existing turntables, or any turning problems that might have been encountered at depots. Problems of some other kind were inevitable. These particularly concerned the trailing truck, with its inside axleboxes set directly beneath the wide firebox, which therefore were very prone to overheat. In addition to this there were

lubrication problems. It was subject to several later modifications.[6]

However, in other respects No. 111's size and weight did indeed tell against it. The originally intended coupled axle load was 19 tons, but when built this came out at nearly 20½ tons (officially it was 20 tons), thereby restricting the engine to the Paddington–Bristol main line. As a junior draughtsman at Swindon in 1907, Harold Holcroft had been involved in the design of 'The Bear', specifically producing the drawings for the cylinders, which while very similar to those for the 'Stars' were increased in bore from 14¼ to 15in, although he would have liked to have been able to increase these to 16in. More than fifty years later he recalled that at the time he had not understood why the engine was built.[7, 8] In 1906, in an early report to the GWR directors, Churchward stated that that there was no special need for such a large locomotive for the time being, and that it was essentially of 'an experimental nature'.

The reasonable assumption that the 4-6-2 would run through to Plymouth was never realised, although it is understood that it might have reached Exeter, Newton Abbot and Wolverhampton on exceptional and unfortunately unrecorded occasions. Its operation might have been unnecessarily restricted in that, remarkably unknown to Churchward while he was in office, the civil engineer had been steadily strengthening bridges on major GWR routes to take heavier locomotives with axle loads of up to 22½ tons. It has been suggested that Paddington–Penzance and even Paddington–Fishguard *non-stop* workings might have been envisaged for it, or any possible successors, given the GWR's extensive resort to carriage 'slipping' at speed without the need to stop the entire train. Around 1908 the GWR had expended huge sums of money in the (ultimately vain) hope of

[6] Wrottesley, M., 'The Great Bear', *Back Track*, September 2008, pp. 547–52.

[7] Holcroft, H., *An Outline of Great Western Locomotive Practice, 1837–1947*, The Locomotive Locomotive Publishing Company Ltd, 1957, p. 87.

[8] Holcroft, H., *Locomotive Adventure*, Ian Allan Ltd, 1962, p. 60.

developing Fishguard as a major transatlantic port.[9] Churchward subsequently showed no apparent enthusiasm for the locomotive, but if nothing else, *The Great Bear* was a positive boon to the GWR publicity department. Normally, having run down to Bristol during the day, it often returned to London by night on a goods working.

When new, on a special working in June 1908, No. 111 hauled a 103-wagon goods train aggregating 1,200 tons from Paddington to Stoke Gifford. On another occasion in 1909 it is said to have worked a goods train of no less than 2,375 tons non-stop between Swindon and Acton at an average speed of 24½mph. Strangely, neither of these runs would appear to have involved the GWR dynamometer car.

For fourteen years until early 1922, GWR No. 111 remained the only British 4-6-2 tender engine. By this time its copper inner firebox was becoming due for replacement. Instead, No. 111 was effectively withdrawn from service in January 1924 with an official 527,272 miles to its credit, and officially rebuilt as a Castle 4-6-0. Although this engine retained the same distinctive running number, it was nevertheless renamed *Viscount Churchill*, after the current GWR chairman. This ran until 1953 with an official estimated aggregate mileage of very nearly 2 million to its credit. Initially retaining the unique bogie tender, that later operated for a number of years behind a number of different 4-4-0 and 4-6-0 GWR passenger locomotives.

The GWR never returned to the Pacific type. Fred Hawksworth, who as a young draughtsman had actually drawn out the general arrangement drawing for *The Great Bear* in 1908, later consistently denied, including to the present author, persistent rumours that when CME of the GWR he had proposed a new 4-6-2 immediately after the Second World War. That said, a very preliminary diagram for such is nevertheless known to have been prepared in Swindon drawing office in 1946, together with a few calculations.

## The Paget Locomotive

An interested observer at the time must have been puzzled by the simple locomotive inventory, sorted by wheel arrangement, which was contained in the Midland Railway section of the 1914 to 1920 editions of *The Railway Year Book*. On a railway dominated by modestly proportioned 4-2-2s, 2-4-0s, 4-4-0s and 0-6-0s, it nevertheless also listed a solitary and extremely mysterious 2-6-2 tender engine, concerning which absolutely nothing had ever been made public. (Ironically, by 1914 this locomotive was no longer in operation, never having entered regular service.)

A photograph of Midland Railway No. 2299 in works grey, together with a remarkably detailed technical description, written by the locomotive historian E. L. Ahrons shortly before his death, was published posthumously in 1927.[10] The rationale of the locomotive's unorthodox design characteristics, accompanied by numerous working drawings, was finally unveiled twenty years later still by James Clayton, not long before his death in late 1946.[11]

The only *official* photograph of Midland Railway eight-cylinder 2-6-2 No. 2299. *Former Ian Allan Archive*

[9] Rutherford, M., A century of Pacific locomotives, Part 2, *Back Track*, February 2002, pp. 64–73.
[10] Ahrons, E. L., *The British Steam Railway Locomotive, 1825–1925*, The Locomotive Publishing Company, 1927, pp. 343–5.
[11] Clayton, J., The 'Paget' Locomotive, *The Railway Gazette*, 2 November 1945, pp. 444–51.

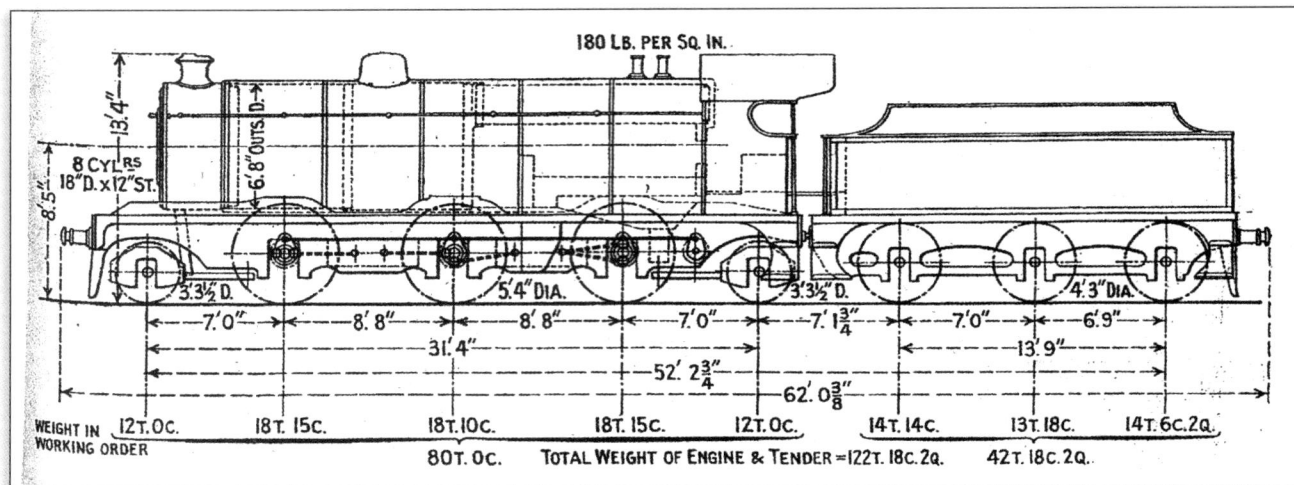

Paget 2-6-2 diagram.

Around 1904, just seventy-five years after the construction of *Rocket*, Cecil Paget, then works manager at the Derby locomotive works of the Midland Railway, whose father, Sir Ernest Paget, rather fortuitously happened to be none other than that company's chairman, seriously began to 'think outside the box' as to how the ubiquitous and now well-established conventional steam locomotive might be further improved upon. His initial thoughts were in terms of a very large outside-frame 4-6-0, a proposal that was later discarded in favour of an alternative 2-6-2 of similar size and features. An exceptional feature would be the employment of no fewer than *eight* cylinders, which would be cast in two groups of four, interposed between the coupled axles and located inside the frames.

The cylinders, however, were to be 'single-acting', inspired by the contemporary Willans high-speed steam engines then employed to drive electricity-generating sets. A very attractive characteristic of these was the perfect balance and the minimal wear of the few moving parts, which resulted from the elimination of crossheads and slide bars. The steam distribution to the cylinders was to be effected by a new form of bronze rotary valve, for which Paget was granted two patents.[12] Having a maximum diameter of almost 7ft, the boiler for its part was to be no less unorthodox, at a time when the conventional locomotive firebox with its hundreds of stays was by far the most problematical portion of a steam locomotive to maintain. Therefore, in order to reduce repair costs stays were to be largely eliminated by cloaking with firebrick the interior of the firebox, which contained no copper, albeit

at the heavy price of largely sacrificing very valuable primary evaporative heating surface. The grate area was no less than 55 sq ft, to service which *two* fireholes, situated side by side Belgian-fashion, were provided.

In order to prepare the working drawings for his brainchild, Paget engaged the services of James Clayton (see Chapter 15), late of Beyer, Peacock & Co. and subsequently with the SECR at Ashford. He had recently, however, defected to the nascent motor car industry in Coventry in 1902, from where Clayton was recruited directly by Cecil Paget. The construction of the unusual locomotive was later sanctioned by the Midland directors, but only after Paget had been promoted from also being Richard Deeley's assistant to become general superintendent of the Midland Railway in April 1907. However, he was still to pay for it out of his own pocket, and it was to be constructed under his direct supervision, despite the fact that he was no longer works manager. In the event, Paget's personal resources failed to cover the full cost of the engine, which required a £2,000 contribution from the Midland Railway in order to complete it. Numbered 2299, the engine emerged from Derby Works in February 1909, when only a single *official* photograph of it was ever taken, and the engine was continually surrounded by extreme secrecy.[13]

Clayton revealed that the 2-6-2 had been designed to haul both heavy goods trains at 15mph, and express passenger trains at over 80mph, a high speed that it evidently could comfortably achieve. Nothing is on record regarding its actual trials, which are said to have been undertaken on Sundays, when reputedly No. 2299 got as

---

[12] Paget, C. W., BP 23,714/1904 (cylinder patent). BP 14,488/1905 (rotary valve).

[13] The original official negative, now held by the National Railway Museum, actually exists in two versions. The regularly published form shows the engine in shop grey with the background blocked out in the traditional manner. The other, as taken, shows the engine coupled to an old Metropolitan-type 4-4-0T, and with a very faint pall of smoke issuing from the chimney, indicating that No. 2299 was in steam. An unnamed apprentice also took two views of the engine, when also in grey primer, with a Box Brownie camera, one of which showed the previously unseen left-hand side, with the main steam pipe running outside the boiler from the dome. Both of these photographs were reproduced in the *Midland Railway Society Journal*, No. 63, Winter 2016, p. 28.

An extremely rare photograph, taken by an unknown Derby apprentice, of the other (left-hand) side of the Paget locomotive, also when it was painted in grey primer. *Courtesy, The Midland Railway Study Centre, Derby*

far afield as London and Manchester. Surprisingly, even Clayton did not provide any precise details on this score, other than asserting that one Sunday at Syston, a little south of Leicester, differential thermal expansion caused the valve assembly to seize at 70mph and the engine consequently blocked the main line for seven to eight hours. He also stated that as a consequence the engine never ran again, but both these assertions were refuted by an unattributed footnote, which must almost certainly have been provided by Kenneth Leech, whose personal memoir of Cecil Paget immediately followed Clayton's article.[14] It is reasonable to assume that a more accurate account of the Syston incident had been related by Paget to Leech when they had encountered each other while both were serving in France on the Western Front in 1915. This maintained that the engine had merely stalled on curved track, furthermore without blocking the main line at all. When another engine was sent to provide assistance, in the course of moving No. 2299, in the absence of active lubrication, this caused the cylinders to crack. These were, however, evidently repaired and the engine proceeded to run numerous further tests, of which regrettably no details whatsoever now survive.

Painted black, the engine had, however, already been abandoned by 1913 and was stored in the paint shop at Derby Works, concealed beneath tarpaulins. At that time it was covertly examined in some detail by Kenneth Leech,

when he was a serving ex-LTSR premium apprentice, during the course of a visit to Midland head office. No. 2299 continued to languish at Derby until May 1920, when it was finally broken up, shortly after Cecil Paget, after his return to Derby following distinguished wartime military service, had resigned as general superintendent of the Midland Railway. After years of intense secrecy, its demise was promptly reported from Derby by one F. G. Carrier, in a short-lived railway periodical.[15] The Paget patent circular valves, although disappointingly nothing else, were for some reason then specially officially photographed before their disposal. No. 2299's hitherto little-used Deeley tender would then see thirty-odd years of further service behind Derby Compound and Class 2 4-4-0s, before being condemned in August 1951. It is possible that the 2-6-2's chimney had already been bequeathed in 1919 to the new 0-10-0 banking engine No. 2290 then under construction. (The respective Derby drawing schedules for MR Nos 2299 and 2290 indicated that their chimneys were to be cast according to the same working drawing, although the chimney as fitted to No. 2290 lacked a *capuchon*.)

A very full account of the Paget locomotive is given in an excellent history of Derby Works by Brian Radford.[16] Although that author spent his entire working life at Derby Works, he made no allusion to the possible survival there of the engine's boiler after 1920. However, in May 2002 a correspondent to *The Model Engineer* recalled taking part

---

[14] Leech, K. H., Midland Railway 8-cylinder 2-6-2 No. 2299 with a note on her designer. *The Railway Gazette*, 2 November 1945, pp. 451–3.

[15] *Locomotive News & Railway Notes*, 10 May 1920, p.137. This periodical was only published between 1919 and 1923. Nearly thirty years later, Frank Carrier would distinctively style the forthcoming British Railways Standard steam locomotives. In 1951 he also very skilfully schemed the BR 9F 2-10-0 as a viable alternative to the 2-8-2 originally proposed, but sadly died in 1952 without seeing the 2-10-0 come to its highly successful fruition two years later.

[16] Radford, J. B., *Derby Works and Midland Locomotives*, Ian Allan, 1971.

in an organised visit to Derby Works one Sunday in November 1957, during which the group was apparently informed that within the works there still remained the boiler from an experimental locomotive that had eight cylinders! This was by no means outside the bounds of possibility, given that the unique high-pressure water tube boiler built in 1929 for the experimental Gresley LNER compound 4-6-4 No. 10,000 functioned at Darlington Works in a stationary capacity from 1937 until 1965.

Tradition had it that there had been bad blood between Cecil Paget and Richard Deeley on account of No. 2299, which led directly to Deeley's resignation. However, while already works manager, in November 1905 Paget was additionally promoted to become chief assistant to Deeley, and in 1906 he married into the aristocracy. In 1907, when further elevated to become general superintendent of the Midland Railway, he became Deeley's superior. It was around then, doubtless assisted by his parentage, that the go-ahead was given to build his special locomotive. Deeley indeed resigned in August 1909, shortly after the engine had appeared, but he had indicated this intention a year earlier when informed of a proposed major reorganisation of the Midland Railway locomotive department, initiated by the general manager, Guy Granet, with which he respectfully did not agree. Nearly fifty years later, Deeley's nephew revealed that the family still possessed a friendly New Year greeting dated 31 December 1909, that had accompanied a small gift sent by Paget to his uncle, who by that time was living in Twickenham. Deeley's brother also recalled that the two men had remained on good terms.[17]

Cecil Paget succeeded to the family baronetcy in 1923, and became the managing director of Steel, Peech & Tozer Ltd, the Sheffield steel manufacturers. He died in 1936 at the age of 62.

## The Lickey Banker

The Midland Railway boasted the most steeply graded stretch of *main line* railway in Britain, the Lickey incline, which consisted of 2 miles graded at 1 in 38 between Bromsgrove and Blackwell, on its route between Bristol and Birmingham. Trains required banking assistance here, for which a special 0-6-0T had been built as early as 1845, and indeed standard 0-6-0 tank engines routinely performed this duty in multiple into the 20th century, as trains became heavier. This was an expensive option and around 1907 thought began to be given to producing a special more powerful locomotive that could be manned by a single engine crew. Over the next five years at least twelve options were examined (opposite):

After two years of further deliberation, the 0-10-0 tender engine option was finally selected, although simplified with the provision instead of a parallel boiler, which was finally authorised in May 1914 and construction was then immediately put in hand. The 0-10-0 had first appeared in Austria in 1900, and was adopted in Italy in 1907 in four-cylinder compound form (FS Class 470). No fewer than 143 of these were built up to 1911, including for service on the line between Turin and Genoa that incorporated a section that was considerably even more steeply graded at 1 in 28½ than the Lickey incline on the

Midland Railway four-cylinder 0-10-0 No. 2290, as finally completed at Derby Works in December 1919. *Former Ian Allan Archive*

17 E. M. Deeley, Letter, *Journal of the Stephenson Locomotive Society*, March–April 1956, p. 88.

## Table 29 Midland Railway banking locomotive proposals, 1907–12

| Date | Type | Notes |
| --- | --- | --- |
| July 1907 | 2-8-2T | Four-cylinder compound |
| May 1910 | 0-10-0T | Taper boiler, non-superheated |
| March 1911 | 2-10-0T | Ditto, but superheated |
| April 1911 | 2-6-6-2T | Fairlie type articulated |
| April 1911 | 0-6-0T+0-6-0T | Two 0-6-0Ts cab to cab |
| July 1911 | 2-6-6-2T | Fairlie type articulated |
| August 1911 | 0-6-0+0-6-0T | Large conventional taper boiler |
| ditto | 0-6-0+0-6-0T | Double-ended boiler with central firebox |
| ditto | 0-6-0T+0-6-0T | Cab to cab, superheated |
| September 1911 | 0-6-6-0T | Rigid frame, four cylinders, superheated taper boiler |
| January 1912 | 2-8-2T | 1907 proposal (above) revived |
| February 1912 | 0-10-0 tender | Four cylinders, superheated *taper* boiler (there appears to have been no preliminary diagram for the engine as actually built with parallel boiler). |

Midland Railway, although this route would very soon be electrified. Quite apart from originally being designed to run cab first (with a coal bunker attached in front) and trailing a water tender, this altogether unusual design was arranged on the Italian Plancher system. Hereby the high-pressure cylinders were cast together as a pair and shared a single piston valve, being mounted on the right-hand side, while the low-pressure cylinders were arranged similarly on the left-hand side, with all four cylinders driving onto the middle coupled axle. A very similar arrangement but involving only simple expansion was followed by Derby on its 0-10-0, whereby one 10in diameter piston valve served both adjacent steeply inclined inside and outside cylinders via cross ports as there was no room beneath the large smokebox to accommodate separate piston valves for the inside cylinders. For example, the front piston valve head served both the front port of an outside cylinder and the back port of the adjacent inside cylinder. This arrangement also saved a considerable number of moving parts. The Walschaerts valve gear, set for outside admission, was interchangeable with that fitted to the Somerset & Dorset 2-8-0s that had just been built, and which had very much been the work of James Clayton, when assistant chief draughtsman. Significantly, a full set of drawings for the Italian 0-10-0s is reported once to have been held at Derby, which in 1906 had sold fifty elderly Kirtley double-framed 0-6-0s to the newly formed Italian State Railways.

At some point after the declaration of war in August 1914 construction of the 0-10-0 at Derby was suspended, but was resumed five years later in 1919 when the engine was officially completed on the final day of that year. Numbered 2290, it made its trial trip on 1 January 1920, almost certainly out to the former isolated triangular station at Trent, and return. Unlike 2-6-2 No. 2299, its construction had been officially extensively recorded photographically in considerable detail, and the engine was widely publicised with the publication of the general

arrangement drawings. The boiler was the largest ever constructed at Derby, which was followed by a second slightly modified unit in 1922. This had been built in order to minimise the time the engine would spend away in works at Derby while undergoing heavy repairs.

On 13 July 1924 No. 2290 made a solitary test run with the ex-LYR dynamometer car between Wellingborough and Brent, at a time when the newly formed LMSR considered that a powerful heavy goods engine was urgently required in order to obviate costly double heading with pairs of 0-6-0s on heavy London-bound coal trains, and was also even considering reviving the LYR 1914 2-10-0 proposal. The 0-10-0 proved to be most unsuitable, however.[18] The locomotive had been specifically designed to develop short bursts of high power at very low speeds, rather like the Great Central 0-8-4Ts, while its cylinder and valve gear arrangements were not conducive to main line haulage. Uniquely soon fitted with a headlight for night-time working, it was nevertheless one of Britain's most successful 'one off' locomotives. It was always stationed at Bromsgrove and remained there in active use until withdrawn from service in May 1956, with a purely theoretical official life mileage of 835,831 to its credit. The author retains from the previous year a still very clear childhood memory of travelling home from a family holiday in South Devon, during the course of which his train benefited from the 0-10-0's extremely vociferous assistance.

One of the two unusual binary cylinder castings (which although inclined were interchangeable with each other) had only fairly recently been replaced, and there were rumours that one of these was to be preserved. Very regrettably this did not prove to be the case. Although the engine was broken up in April 1957, one of these castings was photographed still lying out in the open at Derby Works about three years later. Ironically, when steam banking assistance finally ended on the Lickey Incline in

[18] Cox, E. S., *Locomotive Panorama*, Vol. 1, Ian Allan, 1965, p. 43.

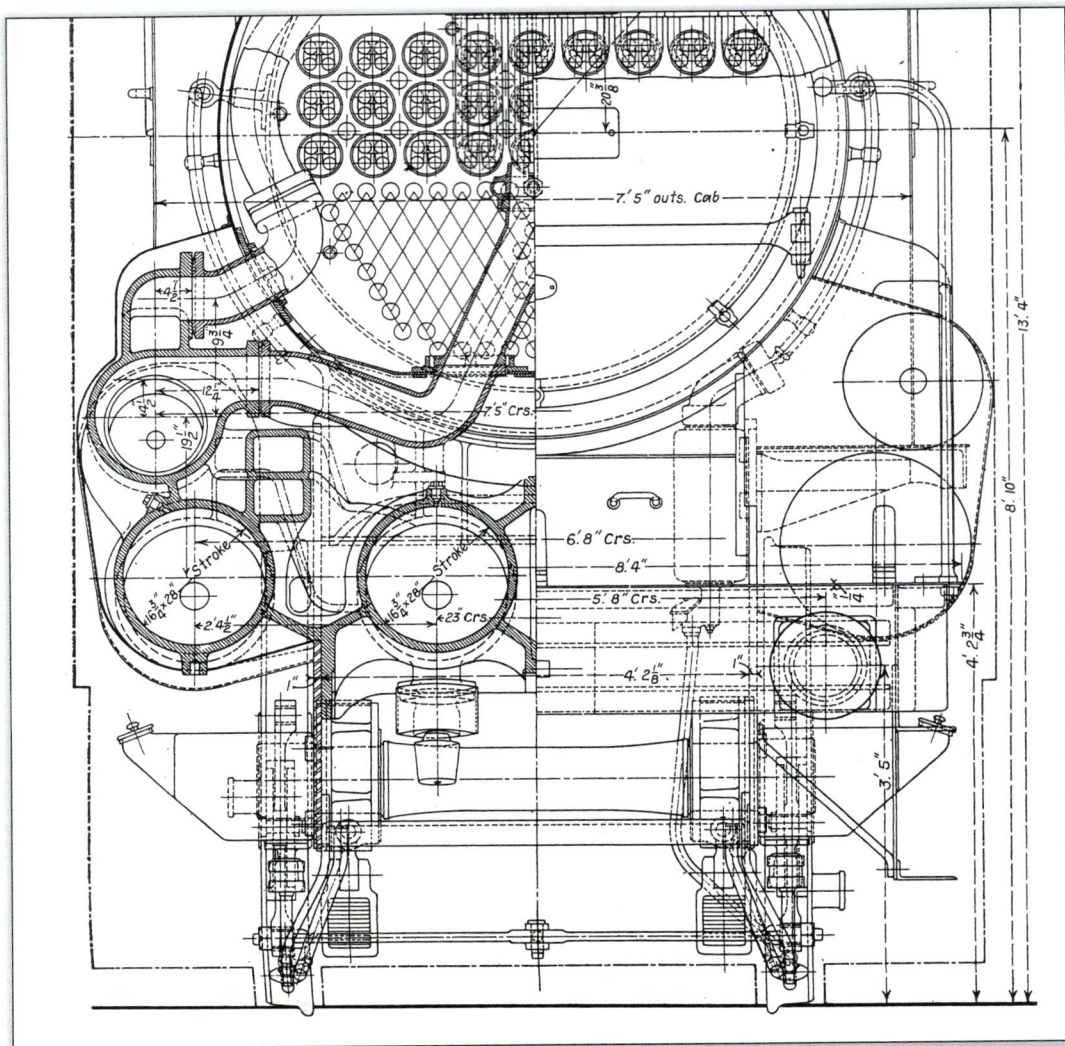

A detail from the sectional general arrangement drawing for Midland Railway 0-10-0 No. 2290 showing the shared piston valve arrangement between adjacent inside and outside cylinders.

1964, this was once again sometimes provided by the combined efforts of up to four 0-6-0 tank engines, which latterly were of Swindon rather than of Derby design.

A fine 1/16 scale model of No. 2290 in LMSR guise, is held by the National Railway Museum, which normally resides on a plinth appropriately inclined at 1 in 38.

## Table 30 Leading dimensions of GER No. 20, GWR No. 111 and MR Nos 2290 & 2299

|  | GER 0-10-0WT No. 20 1902 | GWR 4-6-2 No. 111 1908 | MR 2-6-2 No. 2299 1909 | MR 0-10-0 No. 2290 1919 |
|---|---|---|---|---|
| Cylinders | (3) 18in x 2in | (4) 15in x 26in | (8) 18in x 12in | (4) 16¾in x 28in |
| Driving wheel dia. | 4ft 6in | 6ft 8½in | 5ft 4in | 4ft 7½in |
| Boiler pressure, lb/in² | 200 | 225 | 180 | 180 |
| Evaporative HS ft² | 2873 | 2856 | 2018 | 1718 |
| Superheater | – | 545 | – | 445 |
| Grate area | 42.0 | 41.8 | 55.25 | 31.25 |
| Adhesive weight, tons | 80.0 | 60.0 | 55.7 | 73.65 |
| Engine weight | 80.0 | 97.0 | 74.5 | 73.65 |
| Tractive effort, lb | 39,780 | 27,800 | 18,590 | 43,313 |

# 14
# Edwardian Locomotive Paintwork

Particularly notable aspects of the Edwardian era were the very varied and attractive liveries applied by the different railway companies to their locomotives both large and small. Regrettably this period was too early for these to be recorded in their full glory in colour by the available photographic technology. However, these still shine through even in contemporary monochrome exposures. It is a scientific fact that the human eye can discern a greater number of different shades of green than any other colour, and to which green is also the most restful. A simple survey, below, of the 1901–14 period, reveals that green was by far the most popular basic colour for passenger locomotives as adopted by the various railway companies. It is interesting to note that prior to 1873 the London & North Western Railway had also painted its locomotives green, as did both the Midland and Lancashire & Yorkshire railways before 1883.

### Green (13)
Great Western (chrome), North Eastern (Saxon), Great Central (Brunswick), Great Northern ('grass'), Glasgow & South Western ('dark'), London & South Western, South Eastern & Chatham (Brunswick), Highland, Hull & Barnsley ('invisible'), Great North of Scotland, London Tilbury & Southend (†1912 to MR), Rhymney, Maryport & Carlisle.

### Red (6)
Midland (crimson lake), North Staffordshire (madder lake), Furness (Indian), Barry, Metropolitan (Indian), Midland & South West Junction ('dark').

### Black (5)
London & North Western ('blackberry'), Lancashire & Yorkshire, Taff Vale, Cambrian, Lancashire Derbyshire & East Coast (†1907 to GCR).

### Blue (3)
Caledonian ('sky'), Great Eastern (ultramarine), Somerset & Dorset Joint ('dark').

### Brown (3)
North British (olive), London Brighton & South Coast (from 1906, 'umber'), Midland & Great Northern Joint (ochre).

### Yellow (1)
London Brighton & South Coast (pre-1906, 'daffodil').

† Indicates absorption by another larger railway company (given).

The shades of green employed by the London & South Western, Glasgow & South Western, and Highland are known to have varied over time. The livery employed by the Hull & Barnsley Railway, although *ostensibly* black, was officially termed 'invisible green, appearing black with a green cast in sunlight'. Interestingly its *styling* derived directly from the locomotive livery of the London Chatham & Dover Railway, whose locomotive superintendent, William Kirtley (a nephew of Matthew Kirtley), as a consultant had also been responsible for the first locomotives built for the HBR c. 1885, before the appointment of Matthew Stirling as its locomotive superintendent. The actual styling of the LCDR (black) livery with its scalloped panelling also provided the direct basis for the elaborate Brunswick green livery adopted by the South Eastern & Chatham Railway after 1899. This was almost certainly devised by Robert Surtees, formerly chief draughtsman of the LCDR, who now on seniority grounds occupied the post on the SECR at Ashford, while remaining fiercely loyal to Longhedge traditions.

The livery of the North British Railway has been described as 'difficult to define', and after 1900 at least varied between olive brown and olive green. It clearly owed its origins to William Stroudley, via his close associations with a youthful Dugald Drummond in Glasgow, Inverness and Brighton, before Drummond returned to Cowlairs once again in 1875.

Although colour plates of locomotives appeared in *The Railway Magazine* during the early 1900s,

The prototype Great Central Railway 4-4-2, No. 192, delivered in December 1903, standing on Neasden shed when brand new. *Former Ian Allan Archive*

undoubtedly superior and more accurate representations were the contemporary so-called 'F. Moore' paintings reproduced in *The Locomotive Magazine*. These were actually skilfully colourised official broadside photographs of contemporary locomotives set against idealised and sometimes improbable backgrounds.

The printed specifications for locomotives to be built by contractors included a paragraph or two stipulating how these should be painted, in terms of base coats, applications of varnish, etc. Colours and lining details were often communicated via sample wooden panels that were also supplied. Some examples of such painting specifications are given at the end of this chapter.

In 1901 several railways made no distinction between passenger and goods locomotives in the application of their liveries. Thus the first batch of North Eastern, and all the Great Northern 0-8-0s were originally turned out in fully lined green, but the NER adopted lined black for goods engines from about 1903, and from 1913 the GNR painted its goods engines, including the new Gresley 2-8-0s, plain dark grey. The early GWR 2-8-0s were finished in fully lined green livery and with copper-capped chimneys until 1914. Many 0-6-0s were painted crimson lake on the Midland Railway until plain black became the order of the day for goods engines from 1910, in the interests of economy. In early 1906 the London Brighton & South Coast Railway began to abandon its unique elaborate yellow passenger livery, termed by Stroudley himself as 'Improved Engine Green'. This had been evolved in the early 1870s by William Stroudley, who may have been subject to a rare ocular condition, and possibly perceived it indeed to be a bright shade of green. The LBSCR then instead adopted a less expensive brown (umber) livery for its passenger

locomotives. The corresponding long-established olive green livery for LBSCR goods engines gave way to black.

Some of the most attractive liveries were abandoned after war began in late 1914, including that of the SECR for example, while the last Great Eastern locomotive to be finished in ultramarine was its second 0-6-2T No. 1001, completed in February 1915. On the other hand, the lighter blue of the Caledonian continued unabated, and was religiously reapplied by the Yorkshire Engine Company in Sheffield when it repaired a total of five members of the McIntosh 908 and 918 4-6-0 classes between 1916 and 1920. The Great Central also appeared to be unaffected, turning out its new express passenger 4-6-0 *Lord Farringdon* in 1917 in lined green, despite having in 1913 finished three of the six new Sir Sam Fay 4-6-0s in lined black goods livery, rather than in passenger green. The LNWR and GWR abandoned lining out their locomotives, although the GWR continued to paint their engines plain chrome green. The LNWR reinstated lining out in October 1921, and the GWR a little later on passenger locomotives only.

# Great Central Railway 4-4-2 Specification, April 1905 (tender details are omitted):

Boiler to receive two coats of oxide of iron before being lagged, one whilst hot.

Clothing Plates, two coats of lead colour inside.

Splashers, Cab, Clothing, and Wheels to receive two coats of lead colour, filled up with white lead mixed with gold size, rubbed down, followed up with two other coats lead colour, sand papered, after which two coats Brunswick green in oil.

Outside Frames, Main Frame above platform near Smoke-box, Sand-boxes, Buffers, and Footstep Plates same as Splashers, &c, except two coats crimson lake in oil instead of green.

Outside Frames, Wheels, Sand-boxes, Buffers, Footstep plates, Splashers, Cab, and Clothing to receive one coat under varnish, picked out with black lined afterwards to receive one more coat under varnish and two coats best finishing body varnish. To be flatted down with pumice stone and horse hair between each coat.

Numbers in gold leaf, shaded with blue, to be placed on Buffer Beam, after first coat of varnish.

Company's Coat of Arms (Transfer of which will be supplied) to be placed on each Leading and Driving Splasher.

The brass Number-plate, fixed on Cab Sides, to be painted black between numbers, and with one-eighth white line, half inch within the raised edge.

Inside of Main Frames, Frame Stay, and Slide-bar Brackets to have two coats lead colour, filled up with white lead mixed with gold size, rubbed down, one coat flesh colour, sand-papered, two coats of vermillion, and three coats hard-drying body varnish.

Outside of Main Frames, Bogie, and Guard Bars to receive two coats lead colour, filled up, rubbed down, one coat of ivory black, and one coat ivory black mixed with varnish, and one coat hard-drying body varnish.

Bodies of Straight Axles to receive one coat white lead and one coat varnish.

Ends of Axles black, lined with white, and varnished.

Smoke-box, Bogie, back of Fire-box, Platforms, Brake Hangers, etc, one coat black and one coat Japan.

Inside Cab, one coat lead primer, filled up, rubbed down, sand-papered, two coats stone colour (to sample), one coat under varnish, and one coat finishing body varnish. To be lined as per sample panel.

Two days to intervene between each of the last three coats of varnish.

Buffer Beams same as Inside Frames, with the addition of being lined to sample panel, and finished same as Clothing.

Brake Pipes to have two coats of approved rubber varnish.

The Paint and Varnish to be obtained from Messrs Docker Brothers or approved makers.

Sample Panels will be provided, and the greatest care must be taken that all the Engines are painted strictly in accordance therewith, and of the same shades throughout.

# South Eastern & Chatham Railway Class L 4-4-0 Specification, October 1913:

Each Engine and Tender is to be painted in the following manner:-

The boiler, before being lagged, to receive one coat of boiled oil and one coat of red oxide: the inside of the tender tank to have two coats of red oxide.

The engine boiler, cab, splashers, to be painted with three coats of Brunswick green, fine lined with yellow lines, wheels to have three coats of Brunswick green. The frames to be painted deep red. The tender and tank to be painted with three coats of Brunswick green, fine lined with yellow, wheels to have three coats of Brunswick green. The frames to be painted deep red. The front of buffer-plates, inside of frame, motion plate and stay in front of firebox to be painted vermillion. The inside of the cab to be grained light oak. The whole to be first filled up and rubbed down perfectly smooth, and finished with four coats of best engine copal varnish, and be properly rubbed down between each coat. The smokebox and chimney to have three coats of Blundell's smoke-box black. Details of panelling, &c., will be supplied. The number of the engine to be placed on the front buffer plate of the engine. The initials of the Company 'S. E. & C. R.' to be placed on the side of the tank, letters and figures to sample. The paint and varnish to be of approved manufacture.

# Furness Railway 4-4-2T Specification, January 1914:

Before any painting is done the steel and ironwork must be cleaned, free from scale and rust. The boiler is to receive one coat of boiled oil when warm, and two coats of Torbay red oxide paint before being lagged. The lagging plates to be painted with one coat of Torbay red oxide paint, and one coat of lead colour inside. Cab, splashers, side tanks, and coal bunker, lagging plates on boiler, outside frames and wheels to have two coats of lead colour, stopped with hard stopping, and fine coats of filling up, not less than 5 hours between each coat, then rubbed down, followed by two coats of lead colour faced with pumice stone after each coat and then to have two coats of Indian red and one of varnish. All the painting on outside of engine to have three coats of the best hard drying body varnish supplied by firms approved by the Locomotive Superintendent. The frames, smokebox, chimney, firebox casing, ashpan, footplate, brakework, etc., to have two coats of lead colour and two coats of Japan black. The colouring, picking out, coat of arms, and fine lining to be the same as the pattern panel sent by this Company. The inside of the frames and cross stays to have two coats of Torbay red oxide, and one coat of oak colour to pattern. Front of buffer plates and buffer casing to be painted vermillion; inside of cab to have two coats of lead colour, two coats of filling up, rubbed down, and then two coats of oak colour to pattern. The inside of the tanks to be thoroughly cleaned and scoured, to have two coats of red lead paint, the bottom, sides and rails of coal bunker to have one coat of lead colour and two coats of Japan black. Gilt letters are to be placed on side tanks, samples of which will be supplied by this Company.

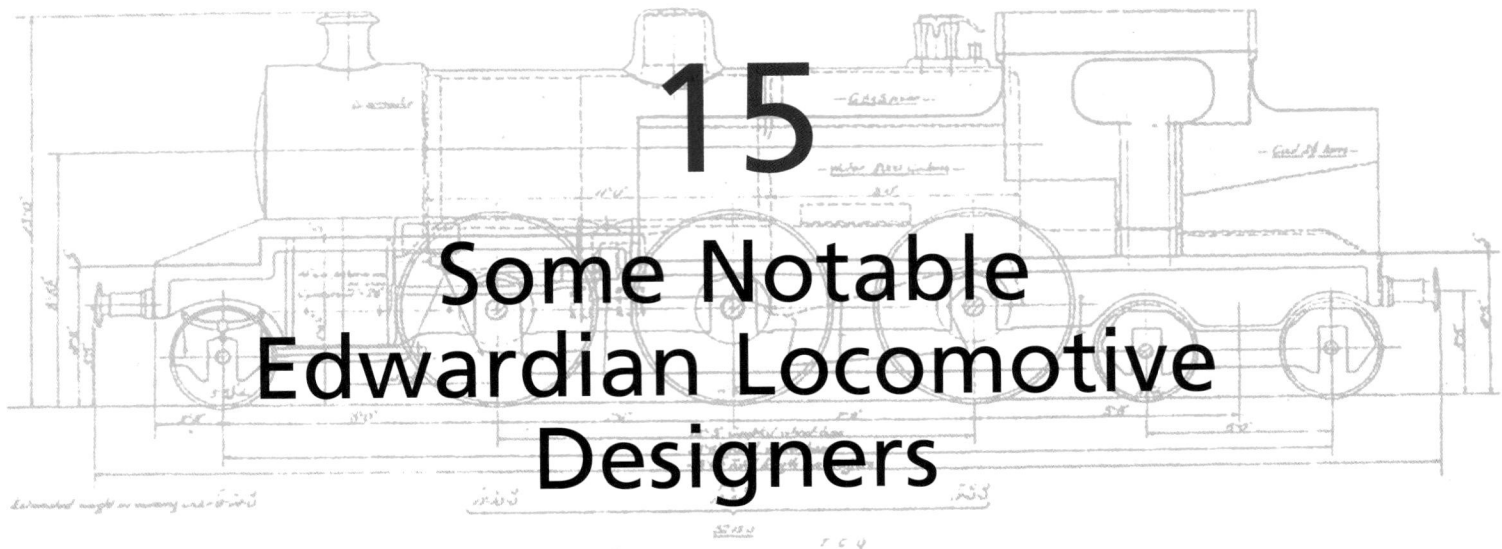

# 15
# Some Notable Edwardian Locomotive Designers

By tradition domestic British locomotive designs were ascribed to the reigning locomotive superintendent or chief mechanical engineer, under whose aegis they appeared. This did not mean, however, that these gentlemen had personally been seated at a drawing board and schemed out a new locomotive type down to the finest detail. They had more wide-ranging and time-consuming administrative duties than this, which on some railways, but by no means all, also encompassed the locomotive running side. Even the chief draughtsman usually no longer worked at a drawing board but interpreted his chief's wishes to the best of his ability, delegating and coordinating the design work within his often quite small team of draughtsmen. Senior among these would be a leading draughtsman, and it was very often he who would work up an initial diagram for provisional approval both in principle by the locomotive superintendent, and then by the chief civil engineer, particularly with reference to estimated weights and physical clearance issues, before more detailed design work could commence.

Many such draughtsmen remain little known to this day, but a few would rise to greater prominence, such as Frederick Hawksworth, the final CME of the Great Western Railway, who alone among his post-1922 contemporaries could indeed truly claim to have actually *designed* steam locomotives at the drawing board. There follows some brief biographical notes concerning a small number of such designers, who were active at the drawing board during the 1901–14 period, and who personally drafted some memorable and sometimes long-lasting designs for Britain's main line railways.

## James Clayton

James Clayton was born in Stockport in 1872 and served a seven-year apprenticeship with Beyer, Peacock & Co. between 1886 and 1893, remaining in the drawing office until joining that at Ashford on the newly established SECR confederation in 1899, where he rejoiced in the nickname of 'Swish'. There he was involved in the design of the new Wainwright Class C 0-6-0 and Class D 4-4-0. Clayton later left Ashford c. 1902 to join the Motor Manufacturing Co. in Coventry as its chief draughtsman. While there he was personally recruited by Cecil Paget, works manager of the Midland Railway at Derby, to prepare privately the working drawings for Paget's proposed revolutionary steam locomotive.

That task completed, in 1907 he was appointed assistant chief locomotive draughtsman at Derby. Seven years later, in early 1914, shortly after having been passed over for promotion to chief draughtsman, he returned to Ashford to enjoy that status under the newly appointed Richard Maunsell. The latter's new 2-6-0 No. 810, when eventually completed at Ashford in 1917, demonstrated unmistakeable Derby features as regards its cab, chimney and tender, and such were later also clearly discernible in the Southern Railway Lord Nelson 4-6-0 built in 1926.[1] In mid-1919 Clayton is believed to have been approached by Beyer, Peacock & Co. to become its general manager, but in order to retain him at Ashford he was instead promoted to become Maunsell's personal assistant.[2] In his later life severely afflicted by arthritis, Clayton retired from the SR in 1938, dying in 1946.

[1] Atkins, P., The James Clayton influence, *Railways South East*, Winter 1988/89, pp. 122–9.

[2] Holcroft, H., *Locomotive Adventure*, Ian Allan Ltd, 1962, pp. 96–7. Holcroft merely referred to 'a private locomotive building works in the north of England'. At his time of writing, Beyer, Peacock & Co., where Clayton had been a former apprentice and draughtsman, was still extant. Significantly, in March 1919 BP secured a contract from the SECR to completely rebuild twenty Wainwright Class E 4-4-0s, a transformation that had been masterminded by James Clayton, which would have brought him into regular contact with his old firm. It is suspected that Richard Maunsell at Ashford successfully nominated his former No. 2 at Inchicore, Edward Watson, instead.

## Basil K. Field

Born in Lee in Kent in 1866, the son of a professor of chemistry, Basil Kingsford Field underwent an unusually extensive education by contemporary standards, including a spell at Heidelberg. He served an apprenticeship at the Ashford Works of the South Eastern Railway under James Stirling, after which he remained in the drawing office. He was reportedly very quickly promoted to become chief draughtsman even before reaching the age of 30. About four years later he suffered demotion when he was obliged to become subservient to Robert Riddle Surtees, late of Longhedge on the London Chatham & Dover Railway, who was by sixteen years his senior, when locomotive design for the newly established SECR combine was unified at Ashford. One suspects that the Ashford drawing office was not a happy place thereafter, in view of the departure of several SER staff for the North Staffordshire Railway.

In mid-1902 Field obtained the post of works manager at Stoke on the NSR. For reasons unknown, but possibly in connection with his forthcoming marriage, he relinquished this post only a few months later, yet temporarily stayed on in some capacity before moving to Coventry as works manager of the Motor Manufacturing Company at the invitation of James Clayton. For the unusual reasons already given, Clayton soon moved back to locomotive work, whereas Field remained with the firm almost until the time it closed down in 1908, following its latter-day removal from Coventry to Clapham in south London. Field became chief locomotive draughtsman on the LBSCR at Brighton in 1907, later becoming promoted to works manager in 1912. He took early retirement following the formation of the Southern Railway in 1923, thereafter devoting much of his time to building model steam locomotives at his home in Brighton.[3, 4] Noted by his friends for his sense of humour and his skill as a raconteur, Field died in Brighton in 1941.

While at Ashford, Field had developed an elegant chimney style for James Stirling, which enjoyed only a relatively short currency on the SER/SECR. It was, however, also later adopted on the NSR, and a little later still (modified with a very slight cap) on the LBSCR.[5] On the locomotives of both of these railways it was much longer lasting; on the latter it could still be seen even until as late as the early 1960s.

## George Heppell

George Heppell was born at Killingworth in 1851, where his family had had close associations with George Stephenson, and he was initially employed an office boy by Robert Stephenson & Co. in Newcastle in 1864. A casual freehand sketch he had made of a locomotive was noted by higher authority, which led to him being taken directly into the drawing office, although his request to serve a proper apprenticeship in the works went unheeded. Much later, for domestic reasons dissatisfied with the pay at Stephenson & Co., in September 1882 he briefly transferred to R & W Hawthorn & Co. next door, before moving in July 1883 to the North Eastern Railway locomotive drawing office in Gateshead directly across the River Tyne, during the brief period when Alexander McDonnell was locomotive superintendent. Of strong personality, for more than twenty years he worked under the chief draughtsman, Walter Mackersie Smith, with whom Heppell had a difficult relationship, but for whom he claimed to have solved a number of major design problems. After 1900 Heppell frequently functioned as acting chief draughtsman on account of Smith's periodic sustained absences through illness, finally achieving his long-held ambition of becoming chief following Smith's death in harness in October 1906. Heppell's twin 'trademarks' were 5ft 6in diameter boilers, and later three cylinders cast together as a single unit or 'monobloc', of which he was the British pioneer. He retired in 1919, and a few years later wrote his private memoirs that, minus a few unexplained excisions that must have been made many years before, were finally published in 2012.[6] These provide a rare insight into the locomotive design process on a major British railway during the late Victorian and Edwardian periods. George Heppell died in Northallerton in 1934.

## Frederick W. Hawksworth

Frederick William Hawksworth was born in 1882 in Swindon, where his father was in the drawing office, which he himself entered in 1905 after serving an apprenticeship in the Works. Three years later he had the honour of preparing the general arrangement drawings for *The Great Bear*, and during 1910–11 was largely responsible for the detailed design of the Churchward 43XX 2-6-0 and 42XX 2-8-0T, both of which would enjoy an overall operating span of more than fifty years. He also schemed the unique GWR small 4-4-2T No. 4600 that was built in 1913.[7] In 1925 Hawksworth was promoted to chief draughtsman, and in 1941 finally became CME following the long overdue retirement of Charles Collett. He died in Swindon in 1976.

[3] Atkins, P., Ashford to Brighton (via Stoke), Basil K. Field – locomotive engineer, *The Southern Way* No. 17, January 2012, pp. 42–7.

[4] The above article was revised and extended with M. G. Fell, and published in *North Staffordshire Railway Study Group Journal* No. 31, October 2012, pp. 19–27.

[5] The Field style of chimney also appears to have been copied by Kerr, Stuart & Co., the commercial locomotive builders, who were also located in Stoke.

[6] Heppell, G., *North Eastern Locomotives: a draughtsman's life*, North Eastern Railway Association, 2012.

[7] Atkins, P., Before they were famous, *Back Track*, December 2013, pp. 725–7.

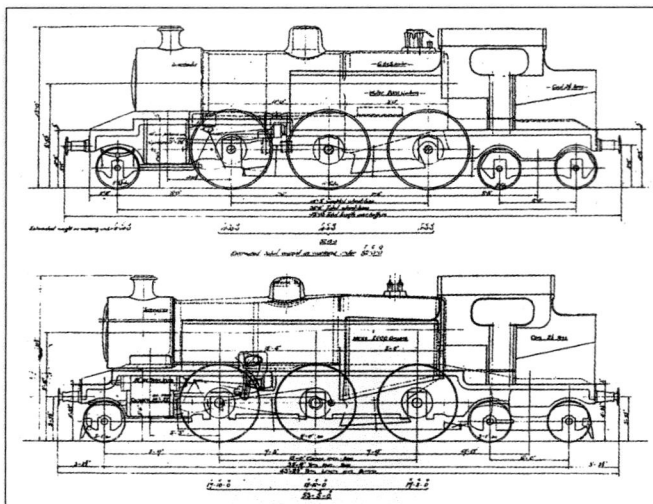

*Upper:* The February 1914 Derby proposal for a 2-6-4T to work on the recently acquired Tilbury section of the Midland Railway in place of the unwanted Whitelegg 4-6-4Ts.

*Lower:* The reworking of the Derby 2-6-4T proposal made by William Hooley at Ashford in May 1914 for the South Eastern & Chatham Railway. This became SECR Class K, of which the prototype No. 790 was completed three years later in July 1917.

## William G. Hooley

Born in Timperley, Cheshire, in 1887, William Glynn Hooley served as an apprentice at Beyer, Peacock & Co. from c 1904, and then stayed on in the drawing office there until July 1911. During that time he executed the general arrangement drawing for the two pioneer 2ft gauge compound 0-4-0 + 0-4-0 Beyer Garratts for Tasmania built in 1909, but designed by his senior colleague Samuel Jackson. (Of these, No. K1 is now preserved on the Ffestiniog Railway in North Wales.) Hooley then moved to the smaller yet longer-established locomotive builder, Nasmyth Wilson & Co. in Patricroft, before securing the post of senior draughtsman on the South Eastern & Chatham Railway at Ashford in January 1913, at the beginning of the end of the Wainwright regime there.

Under Wainwright's successor, Richard Maunsell, in mid-1914 Hooley produced the initial schemes for the Class N 2-6-0 and Class K 2-6-4T. He then worked on proposed national standard 2-6-0 and 2-8-0 schemes (derived from the N class 2-6-0) for the ARLE. Shortly after the end of the First World War in May 1919, Hooley produced outline schemes for proposed 2-8-0 tank engines derived from the Class N, alternatively having either two or three cylinders, although in the event neither of these was built. In 1920 Hooley undertook the detailed design work for a three-cylinder version of the Class N 2-6-0, which as SECR No. 822 was completed at Ashford Works in December 1922. This incorporated Harold Holcroft's conjugated valve gear to operate the middle piston valve.[8] In his memoirs, Holcroft, ex-Swindon and about the same age as Hooley, appeared to claim for himself the credit for designing No. 822 single-handedly, despite the fact that it had been Hooley, for instance, who had made the complex full-scale drawings for its cylinders.[9, 10]

Following the formation of the Southern Railway in 1923, Hooley was promoted to leading draughtsman, working from a small drawing office at Waterloo Station. There he evolved the preliminary schemes, sometimes several years in advance, for most of the SR locomotives classes ascribed to Richard Maunsell, most notably the Lord Nelson four-cylinder 4-6-0, and Schools three-cylinder 4-4-0, both of which were subsequently somewhat 'watered down' by Thomas Finlayson, the chief draughtsman at Eastleigh, prior to their construction there. Hooley also designed the Class Z three-cylinder heavy-shunting 0-8-0T. In 1935 he began to suffer serious health problems and died suddenly in January 1936 at the early age of 48.[11]

## Carl H. Schobelt

Born in Chorlton, Manchester, in 1863, the son of an immigrant German instrument maker, Carl Heinrich Schobelt served his time at Sharp, Stewart & Co. at the Atlas Works in central Manchester between 1878 and 1883, staying on as a draughtsman there until 1888 when Sharp & Co. removed to Glasgow. Schobelt then transferred to Beyer, Peacock & Co., where he eventually became chief locomotive designer in 1902. The following year he schemed the prototype alternative 4-4-2 and 4-6-0 express passenger locomotives for the neighbouring Great Central Railway.[12] (Interestingly, at this time Carl Schobelt and John Robinson both resided in the Heatons district of Stockport, and would also doubtless have been professionally well-acquainted in view of the close proximity of their respective workplaces.) Later, in 1907 Schobelt also schemed the three-cylinder 0-8-4T heavy-shunting engine for the GCR. Around 1918 he became assistant to the general manager, and in 1925 chief draughtsman. He died in 1927 shortly before he had been due to retire.

[8] Holcroft, H., BP 7859/1909 (locomotive valve gear).

[9] Holcroft, H., *Locomotive Adventure*, Ian Allan Ltd, 1962, pp. 93–6.

[10] Atkins, P., (W. G. Hooley) The man who *really* designed the Maunsell 'Moguls', *The Southern Way*, January 2011, pp. 16–23.

[11] *The Locomotive*, February 1936, p. 63, W. G. Hooley obituary.

[12] Atkins, P., Robinson Great Central locomotives, a postscript, *Back Track*, October 2010, pp. 634–5. Schobelt's 4-4-2/4-6-0 and 0-8-4T schemes for the GCR survive in the Beyer Peacock archive held by the Museum of Science & Industry in Manchester.

# 16
# Edwardian Locomotive Nicknames

The dramatic increase in size of British locomotives, particularly after 1900, gave rise to some gaining nicknames, some appropriate or ironic. Others simply reflected contemporary events, often of a military nature, most notably the Second South African War (1899–1902), but also personalities and even dance crazes.

*Austrian Goods*
> GSWR 403 class 2-6-0, 1915
> Nickname applied for unknown reasons. It is very unlikely that the builder (NBL) had built these locomotives using materials obtained for an alleged locomotive order originally received from Austria–Hungary, (and subsequently cancelled) as has sometimes been suggested.

*Bill Bailey*
> LNWR 1400 class compound 4-6-0, 1903
> *Bill Bailey, Won't You Please Come Home?* was a popular song first published in the USA in 1902.

*Crab*
> GCR 1B class 2-6-4T, 1914
> When in operation these locomotives suggested a sideways 'crab-like' motion.

*Crystal Palace*
> GER 1300 class 2-4-2T, 1909
> Diminutive branch line passenger tank engine with conspicuously large side window cab.

*Decapod*
> GER 0-10-0WT, 1902
> 'Decapod', literally meaning 'ten-footed', was a term already applied to ten- coupled (2-10-0) locomotives in the USA.

*Dreadnought*
> LYR 1506 class 4-6-0, 1908
> HMS *Dreadnought* was a revolutionary all big gun, turbine-powered battleship, completed by the Royal Navy at Portsmouth in 1906. This precipitated a naval arms race with Imperial Germany that reached its peak

A 'Sea Pig', a Lancashire & Yorkshire Railway 0-8-0 with 'Cornish'-type boiler. *Former Ian Allan Archive*

in 1908, coincident with the appearance of these massive LYR 4-6-0s.

### Flatiron

MR 0-6-4T, 1907

An unusual feature of these locomotives was the extension of their side tanks to the front of the smokebox, thereby requiring a large aperture to permit access to the inside motion for oiling purposes, so resembling a contemporary domestic iron.

### Italian

Highland Railway River class 4-6-0, 1915 (sold to Caledonian Railway)

This was term was possibly applied because these locomotives changed their ownership around the time that Italy transferred its alliance from the Central Powers to the Allies during the early months of the First World War.

### Jersey Lily

GCR 8B class 4-4-2, 1903

The popular sobriquet of Emilie le Breton (stage name Lily Langtry) (1853–1929) who was a well-known actress of the late Victorian/early Edwardian period, and a former mistress of King Edward VII.

### Kruger

GWR double-framed 4-6-0, 1899, and 2-6-0s, 1901–03

Paul Kruger (1825–1904) was President of the South African Republic between 1883 and 1902, and very much a hate figure in Britain during the South African War of 1899–1902.

### Little Egbert

LYR 0-8-2T, 1908

Little Egberts was reputedly the name of a troupe of circus elephants.

### Long Tom

GNR Q1 class 0-8-0, 1901

The large boiler of this locomotive was suggestive of the French-manufactured 155mm heavy field gun, with 14ft long gun barrel giving a maximum range of 6 miles, employed by the Boer forces during the South African War.

### Paddlebox

LSWR T14 class 4-6-0, 1911

The unusually prominent coupled wheel splashers of these locomotives were suggestive of contemporary paddle steamers.

### Pom Pom

GCR 9J class 0-6-0, 1901

37mm light artillery employed by the Boer forces during 1899–1902, which was also later adopted by the British Army.

### Pumper

GSWR 279 class 0-6-0, 1913

So termed on account of the pumps that were fitted in order to handle the heated feed water when exhaust steam was diverted into the tender.

A GNR 'Long Tom' 0-8-0 is seen at work on a heavy goods train. *Former Ian Allan Archive*

Great Eastern Railway 'Crystal Palace' 2-4-2T. *Former Ian Allan Archive*

*Ragtimer*
  GNR H3 class 2-6-0, 1914
  This reflected the current craze for American jazz music.

*Sea Pig*
  LYR 0-8-0 with 'Cornish'-type boiler, 1903
  Sea Pig was a term sometimes applied to dolphins, which the rather bulbous boilers of these locomotives perhaps suggested.

*Small Hopper*
  LSWR K10 class 4-4-0, 1901
  The origin of this term is unclear. Curiously, the larger-boilered L11 5ft 7in 4-4-0s that immediately followed for their part do not appear to have been called Large Hoppers, however.

*Tango*
  GNR O1 class 2-8-0, 1913
  A dance that originated in Argentina in the 1880s and very much in vogue in Europe in 1913, when the first Gresley 2-8-0s appeared.

*Tiny*
  HBR A class 0-8-0, 1907
  An 'ironic' nickname that simply reflected the much greater size of these locomotives when compared to previous HBR 0-6-0s.

*Togo*
  LYR 2-6-2T, 1903
  Admiral Heihachiro Togo (1848–1934) of the Imperial Japanese Navy was the victor of the Battle of Tsushima (27/28 May 1905) during the Russo-Japanese War.

*Wath Daisy* (or *Daisy*)
  GCR 8H class 0-8-4T, 1907
  Another 'ironic' nickname for a large locomotive, designed specifically for operation at the new Wath concentration yard in South Yorkshire.

*Whitby Willie*
  NER W class 4-6-0T, 1907
  A class of tank locomotive, later rebuilt to 4-6-2T, specifically built for service on the heavily graded coastal line between Scarborough and Whitby, and quite coincidentally designated Class W.

*Yorkie*
  NBR 1 class 4-4-2T, 1911
  This nickname simply reflected the fact that these thirty locomotives had been built by the Yorkshire Engine Company in Sheffield.

# 17

# A Locomotive Chronology, 1901–14

New British locomotive designs introduced or initiated between 1901 and 1914.

**Key to abbreviations:**

*superheated, (*) class only part superheated when first built,

I/C inside cylinders, O/C outside cylinders, 3C three cylinders, 4C four cylinders,

8C eight cylinders, 3CC three-cylinder compound, 4CC four-cylinder compound.

## 1901

| | |
|---|---|
| February | SECR D 4-4-0 (I/C) |
| | HBR F1 0-6-2T (I/C) (as LDECR A class of 1895) |
| | GNR K1 0-8-0 (I/C) |
| April | LSWR E10 4-2-2-0 (4C) |
| May | LNWR Alfred the Great 4-4-0 (4CC) |
| | Met. F 0-6-2T (I/C) |
| June | GWR 2602 2-6-0 (I/C) |
| | LBSCR B4 4-4-0 (I/C) |
| Uncertain | FR 126 4-4-0 (I/C) |
| July | CR 600 0-8-0 (I/C) |
| August | LNWR B 0-8-0 (4CC) |
| | NER T 0-8-0 (O/C) |
| September | GCR 9J 0-6-0 (I/C) |
| October | GCR 11B 4-4-0 (I/C) |
| November | HBR F2 0-6-2T (I/C) |
| December | HBR G3 0-6-0T (I/C) |
| | LSWR K10 4-4-0 (I/C) |

## 1902

| | |
|---|---|
| January | MR 2631 4-4-0 (3CC) |
| | VoRR 2-6-2T (1' 11½in gauge) (O/C) |
| February | GWR 100 4-6-0 (O/C) |
| | TVR O3 0-6-2T (I/C) |
| | RSBR 25 0-6-2T (I/C) (as LDECR A class of 1895) |
| May | CR 55 4-6-0 (I/C) |
| July | GNR No. 271 4-4-2 (4C) |
| September | WLLR 0-6-0T (2ft 6in gauge) (O/C) |
| October | NER U 0-6-2T (I/C) |
| November | GCR 8 4-6-0 (O/C) |
| | GCR 8A 0-8-0 (O/C) |
| | LBSCR E5 0-6-2T (I/C) |
| December | GNR C1 4-4-2 (O/C) |
| | GER 0-10-0WT (3C) |

## 1903

| | |
|---|---|
| January | MR 3 0-6-0 (I/C) |
| March | LNWR 1400 4-6-0 (4CC) |
| | CR 49 4-6-0 (I/C) |
| | GNR R1 0-8-2T (I/C) |
| | GWR Saint 4-6-0 (O/C) (prototype, orig. No. 98) |
| | LYR (Corrugated f'box) 0-8-0 (I/C) (as new build) |
| | GWR City 4-4-0 (I/C) |
| April | Cambrian. 89 0-6-0 (O/C) |
| May | NBR 317 4-4-0 (I/C) |
| | GCR 9K 4-4-2T (I/C) |
| | GSWR 381 4-6-0 (O/C) |
| | LSWR L11 4-4-0 (O/C) |
| June | LSWR S11 4-4-0 (I/C) |
| | GWR 28XX 2-8-0 (prototype, orig. No. 97) |
| | Wirral 14 4-4-4T (I/C) |
| | LTSR 0-6-2T (I/C) |

| | | | |
|---|---|---|---|
| *July* | EWJR 13 2-4-0 (I/C) | *July* | NSR 19 2-4-0 (I/C) |
| *September* | EWJR 14 0-6-0 (I/C) | | GNR 1300 4-4-2 (4CC) |
| | GWR 31XX 2-6-2T (O/C) | *September* | LSWR F13 4-6-0 (4C) |
| | (prototype, orig. No. 99) | | GWR County 4-4-2T (O/C) |
| | GWR (Taper boiler) Bulldog 4-4-0 (I/C) | *October* | MR 4 4-4-0 (3CC) |
| *October* | GWR No. 102 4-4-2 (4CC) | *December* | GCR 8E/8F 4-4-2 (3CC) |
| | (French built) | | LBSCR H1 4-4-2 (O/C) |
| | GWR No. 101 0-4-0T (O/C) | | |

**1906**

| | |
|---|---|
| | GWR (Taper boiler) 36XX 2-4-2T (I/C) |
| | GER Tram 0-6-0T (O/C) |
| | HR 0-6-0T (O/C) |
| | LYR 2-6-2T (I/C) |

| | | | |
|---|---|---|---|
| *November* | CR 492 0-8-0T (I/C) | *February* | SECR E 4-4-0 (I/C) |
| | SDJR 69 4-4-0 (I/C) | *March* | LBSCR C3 0-6-0 (I/C) |
| | NER V 4-4-2 (O/C) | *April* | NER 4CC 4-4-2 (4CC) |
| *December* | GER (Belpaire) Claud Hamilton | | GWR 40 4-4-2 (4C) |
| | 4-4-0 (I/C) | | NER P3 0-6-0 (I/C) |
| | GCR 8B 4-4-2 (O/C) | | GWR 3150 2-6-2T (O/C) |
| | GCR 8C 4-6-0 (O/C) | *May* | GSWR 266 0-4-4T (I/C) |

**1904**

| | |
|---|---|
| | LNWR Precursor 4-4-2T (I/C) |
| | CR 903 4-6-0 (I/C) |
| | NWNGR *Russell* 2-6-2T |
| | (1ft 11½in gauge) (O/C) |

| | | | |
|---|---|---|---|
| *February* | FR L2 0-6-2T (I/C) | | *Superheated* GWR Saint 4-6-0* |
| *March* | LNWR Precursor 4-4-0 (I/C) | | (O/C) (new engine) |
| *May* | LDECR D 0-6-4T (I/C) | *June* | GCR 8F 4-6-0 (O/C) |
| | GWR County 4-4-0 (O/C) | | NBR 848 0-6-0 (I/C) |
| | CR 140 4-4-0 (I/C) | | CMLR 0-6-2T (2ft 3in gauge) (O/C) |
| *June* | LSWR L12 4-4-0 (I/C) | *July* | NBR 868 4-4-2 (O/C) |
| | NER P2 0-6-0 (I/C) | | CR 918 4-6-0 (I/C) |
| | Rhymney. 106 0-6-2T (I/C) | *August* | GCR 5A 0-6-0T (O/C) |
| | N&BR 11 0-6-2T (I/C) | *September* | GCR 8G 4-6-0 (O/C) |
| | LMLR 2-6-2T (2ft 6in gauge) (O/C) | *October* | GWR 45XX 2-6-2T (O/C) |
| *Uncertain* | FR 98 0-6-2T (I/C) | | NBR Intermediate 4-4-0 (I/C) |
| *July* | Cambrian. 94 4-4-0 (I/C) | | CR 908 4-6-0 (I/C) |
| *October* | GWR 44XX 2-6-2T (O/C) | *November* | *Superheated* LYR Aspinall 0-6-0* |
| | GSWR 240 4-4-0 (I/C) | | (I/C) (new engine) |
| *November* | SECR H 0-4-4T (I/C) | *December* | LNWR 19in 4-6-0 (I/C) |
| *December* | NBR 836 0-6-0T (O/C) | | |

**1907**

| | | | |
|---|---|---|---|
| | LBSCR E6 0-6-2T (I/C) | *January* | GSWR 272 0-4-0T (O/C) |
| | MGNJR A 4-4-2T (O/C) | *February* | HBR A 0-8-0 (I/C) |

**1905**

| | |
|---|---|
| | GWR Star 4-6-0 (4C) |

| | | | |
|---|---|---|---|
| *February* | GWR Saint as 4-4-2 (O/C) | *March* | MR 999 4-4-0 (I/C) |
| | (as new build, later reb. to 4-6-0) | | FR 3 0-6-0 (I/C) |
| *March* | GNR No. 292 4-4-2 (4CC) | *April* | MR 2000 0-6-4T (I/C) |
| | HR 0-4-4T (I/C) | | GNR N1 0-6-2T (I/C) |
| *April* | LNWR Experiment 4-6-0 (I/C) | | LYR Compound 0-8-0 (4CC) |
| | GER (Belpaire) 0-6-0 (I/C) | *May* | TVR O4 0-6-2T (I/C) |
| | LYR (Belpaire) 2-4-2T (I/C) | *June* | GCR 9L 4-4-2T (I/C) |
| *May* | Cardiff. 11 0-6-2T (I/C) | | Rhymney. S 0-6-0T (I/C) |
| | KESR 0-8-0T (O/C) | | GSWR 18 4-4-0 (I/C) |
| *June* | GWR 103 4-4-2 (4CC) (French built) | | LBSCR I1 4-4-2T (I/C) |
| | MSWJR 1 4-4-0 (I/C) | | |

| | |
|---|---|
| August | GNR No. 1421 4-4-2 (4CC) |
| | MR 0-4-0T (O/C) |
| October | LBSCR I3 4-4-2T(*) (I/C) |
| November | LSWR E14 4-6-0 (4C) |
| December | NSR M 0-4-4T (I/C) |
| | NER W 4-6-0T (I/C) |
| | GCR 8H 0-8-4T (3C) |
| | LBSCR I2 4-4-2T (I/C) |
| | Rhymney. 0-6-2T (I/C) |
| | PDSWJR 0-6-2T (O/C) |

**1908**

| | |
|---|---|
| February | GWR 111 4-6-2*(4C) |
| March | LYR 0-8-2T (I/C) |
| | LSWR G14 4-6-0 (4C) |
| | MCR 18 0-6-0 (I/C) |
| | (as GSWR 361 class of 1900) |
| May | HR Big Ben 4-4-0 (I/C) |
| | GWR Flower 4-4-0 (I/C) |
| June | LYR 1506 4-6-0 (4C) |
| August | CVHR 0-6-2T (I/C) |
| | GNR Ivatt J21 0-6-0 (I/C) |
| September | LBSCR I4 4-4-2T* (I/C) |
| | NWNGR Fairlie 0-6-4T (1ft 11½in gauge) (O/C) |
| December | SMJR 17 0-6-0 (I/C) |
| | NSR New L 0-6-2T (I/C) |
| | Cardiff. 33 0-6-2T (I/C) |

**1909**

| | |
|---|---|
| January | MR Paget 2-6-2 (8C) |
| February | SECR P 0-6-0T (I/C) |
| April | KER 2-6-0T (O/C) |
| | Superheated GCR 9J 0-6-0* (I/C) (new engine) |
| May | LTSR 79 4-4-2T (O/C) |
| June | HR 0-6-4T (I/C) |
| | GER 1300 2-4-2T (I/C) |
| | NBR Scott 4-4-0 (I/C) |
| July | Rhymney. P 0-6-2T (I/C) |
| September | NER X 4-8-0T (3C) |
| | NBR 239 0-4-4T (I/C) |
| | NBR 858 0-6-2T (I/C) |
| October | Superheated GNR Q2 0-8-0* (I/C) (as new build) |
| December | NSR H 0-6-0 (I/C) |
| | ADR 0-6-2ST (O/C) |

**1910**

| | |
|---|---|
| January | LNWR G 0-8-0 (I/C) (as new build) |
| March? | FR 19 0-6-0T (I/C) |
| May | NER V1 4-4-2 (O/C) |
| | Rhymney. A 0-6-2T (I/C) |
| | GWR 1361 0-6-0ST (O/C) (to Sharp Stewart 1872 design) |
| June | NSR G 4-4-0 (I/C) |
| July | LNWR George V* (I/C) |
| | Queen Mary 4-4-0 (I/C) |
| | LYR Large boiler 0-8-0 (I/C) |
| | GSWR 17 4-4-0 (I/C) |
| | Superheated CR Dunalastair IV 4-4-0* (I/C) (new engine) |
| August | Superheated GNR C1 4-4-2* (O/C) (as new build) |
| October | LSWR P14 4-6-0 (4C) |
| | NER Y 4-6-2T (3C) |
| | Superheated GWR Star 4-6-0* (4C) (as new build) |
| December | LNWR 4-6-2T(*) (I/C) |
| | NSR H1 (Belpaire) 0-6-0 (I/C) |
| | HBR J 4-4-0 (I/C) |
| | GWR 42XX 2-8-0T* (O/C) |
| | LBSCR J1 4-6-2T* (O/C, Stephenson vg) |

**1911**

| | |
|---|---|
| March | GNR D1 4-4-0* (I/C) |
| | LSWR T14 4-6-0 (4C) |
| | Superheated GWR 28XX 2-8-0* (O/C) (as new build) |
| April | Superheated LYR (Belpaire) 2-4-2T (I/C) (as new build) |
| June | GWR 43XX 2-6-0* (O/C) |
| | LBSCR H1 4-4-2* (O/C) |
| | GCR 9N 4-6-2T* (I/C) |
| July | NER Z 4-4-2(*) (3C) |
| | GSWR 128 4-6-0* (O/C) |
| | SMR 0-6-2T (O/C) |
| October | LNWR Prince of Wales 4-6-0* (I/C) |
| | MR.4 0-6-0* (I/C) |
| | Superheated GWR Saint 4-6-0* (O/C) (as new build) |
| November | NSR K 4-4-2T* (I/C) |
| December | LNWR 0-8-2T (I/C) |
| | NER S2 4-6-0(*) (O/C) |
| | HBR L1 0-6-0 (O/C) |
| | GCR 8K 2-8-0* (O/C) |
| | GER 1500 4-6-0* (I/C) |
| | NBR 1 4-4-2T (I/C) |
| | Superheated GWR County 4-4-0 (O/C) (as new build) |

**1912**

| | |
|---|---|
| January | FR 94 0-6-2T (I/C) |
| | CR 498 0-6-0T (O/C) |
| February | LNWR G1 0-8-0* (I/C) |
| | LSWR D15 4-4-0 (I/C) |
| March | LBSCR J2 4-6-2T* (O/C), |
| | (Walschaerts vg) |
| | LYR (Belpaire) 0-6-0* (I/C) |
| June | *Superheated* GWR County 4-4-2T* |
| | (O/C) (as new build) |
| July | CR 30 0-6-0* (I/C) |
| August | GNR H2 2-6-0* (O/C) |
| September | NBR Scott 4-4-0* (I/C) |
| October | NSR KT 4-4-0* (I/C) |
| November | GNR Gresley J21 0-6-0* (I/C) |
| | CR 34 2-6-0* (I/C) |
| | GER E72 0-6-0* (I/C) |
| | *Superheated* LYR Large boiler 0-8-0* |
| | (I/C) (as new build) |
| December | GCR 1 4-6-0* (I/C) |

**1913**

| | |
|---|---|
| January | LNWR Claughton 4-6-0* (4C) |
| | NER Uniflow S2 4-6-0* (O/C) |
| February | NER T2 0-8-0* (O/C) |
| March | FR 130 4-4-0 (I/C) |
| | FR 1 0-6-0 (I/C) |
| April | GSWR 279 0-6-0 (I/C) |
| | (LTSR) 87 4-6-4T* (O/C) |
| | (delivered to MR 12.1912) |
| | GER 0-4-0T (I/C) |
| May | NSR (Belpaire) New L 0-6-2T (I/C) |
| June | GCR 1A 4-6-0* (I/C) |
| | LBSCR E2 0-6-0T (I/C) |
| | GSWR 131 4-4-0 (I/C) |
| August | GCR 11E 4-4-0* (I/C) |
| September | NBR Glen 4-4-0* (I/C) |
| | LBSCR K 2-6-0* (O/C) |
| October | SECR J 0-6-4T* (I/C) |
| | NER D 4-4-4T* (3C) |
| November | GWR 4600 4-4-2T (O/C) |
| | HBR F3 0-6-2T (I/C) |
| December | GNR O1 2-8-0* (O/C) |
| | GNR J23 0-6-0T (I/C) |
| | CR 179 4-6-0* (I/C) |

**1914**

| | |
|---|---|
| February | Wirral. 4 0-4-4T (I/C) |
| | LSWR H15 4-6-0(*) (O/C) |
| | SDJR 80 2-8-0* (O/C) |

| | |
|---|---|
| April | FR 92 0-6-2T (I/C) |
| | LBSCR L 4-6-4T* (O/C) |
| | GNR H3 2-6-0* (O/C) |
| May | SDJR 70 4-4-0* (I/C) |
| | Barry L 0-6-4T (I/C) |
| July | NSR C 0-6-4T* (I/C) |
| | Southwold. 0-6-2T (O/C) |
| | (3ft 0in gauge) |
| August | SECR L 4-4-0* (I/C) |
| October | TVR A 0-6-2T (I/C) |
| December | NBR 8 0-6-0* (I/C) |
| | GCR 1B 2-6-4T* (I/C) |

**'Posthumous'**

Locomotive classes proposed, designed and/or authorised before 1915

| | |
|---|---|
| January 1915 | *1915* GER 0-6-2T (*) (I/C), ordered 4.1914 |
| | GSWR 137 4-4-0* (I/C), ordered pre-1915 |
| | HBR LS 0-6-0* (I/C), ordered 1914 |
| March 1915 | FR 4-4-2T (I/C), specification dated 1.1914 |
| September 1915 | GSWR 403 2-6-0* (I/C), ordered 12.1914 |
| | HR River 4-6-0* (O/C), initial diagram dated 8.1913 |
| | [ordered 9.1914, sold 10.1915, became CR 938 class] |
| November 1915 | Met. G 0-6-4T* (I/C), ordered 9.1914 |
| February 1916 | CR 113 4-4-0* (I/C), proposed 8.1914 |
| November 1916 | CR 60 4-6-0* (O/C), first proposed 2.1911 with 6ft 6in coupled |
| | wheels, modified 6.1914 with 6ft 1in wheels & 6 wheel tender |
| July 1917 | SECR K 2-6-4T* (O/C), initial diagram dated 5.1914 |
| August 1917 | SECR N 2-6-0* (O/C), initial diagram dated 7.1914 |
| June 1918 | NER Uniflow Z 4-4-2* (3C), ordered & designed in 1914 |
| December 1918 | *1919* MR 0-10-0* (4C), ordered 5.1914 |

## Key to abbreviations for smaller railways

| | | | |
|---|---|---|---|
| ADR | Alexandra Docks (South Wales) | MSWJR | Midland & South Western Junction |
| CMLR | Campbeltown & Machrihanish Light (n.g.) | N&BR | Neath & Brecon |
| | | NWNGR | North Wales Narrow Gauge (aka Welsh Highland) (n.g.) |
| EWJR | East & West Junction (later Stratford upon Avon & Midland Junction) | PDSWJR | Plymouth Devonport & South West Junction |
| KESR | Kent & East Sussex | RSBR | Rhondda Valley & Swansea Bay[3] |
| KER | Knott End (formerly Garstang & Knott End) | SDJR | Somerset & Dorset Joint |
| | | VoRR | Vale of Rheidol (n.g.) |
| LDECR | Lancashire Derbyshire & East Coast[1] | WLLR | Welshpool & Llanfair Light (n.g.) |
| LMLR | Leek & Manifold Light (n.g.) | | |
| LTSR | London Tilbury & Southend[2] | [1] Absorbed by Great Central Railway 1907 | |
| Met. | Metropolitan | [2] Absorbed by Midland Railway 1912 | |
| MGNJR | Midland & Great Northern Joint | [3] Operated by GWR from 1906 | |
| | | (n.g.) | narrow gauge |

# Epilogue

With the declaration of war in August 1914 the railways were taken under direct government control, and by the time this phase ended six years later their days as independent entities were numbered. The Railways Act 1921 foreshadowed the amalgamation or grouping of 120 railways into four groups, which legally came into being on 1 January 1923. To the contemporary observer the rapid demise of the former locomotive liveries alone must have been distressing; one former Caledonian engine, a 4-4-0T, remained blue until 1928. Each of the new companies understandably pursued a policy of locomotive standardisation, initially perpetuating recent designs that they had inherited from their constituents, before developing new ones of their own. As a result, locomotive diversity plummeted dramatically, from a grand total of 886* officially recognised locomotive classes in 1923, this almost exactly halved to 'only' 448 by

The last Edwardian. The fifth built of the 120 NER Class T2 0-8-0s produced between 1913 and 1921, and one of the last three to be withdrawn from service in September 1967, is seen as British Railways No. 63344 at Hartlepool c. 1966. *R. J. Carmen*

1948. The latter total actually included about seventy new classes that had been introduced since 1922 (although three such, the two Gresley LNER 2-8-2 classes and the Hughes LMSR 4-6-4Ts, no longer existed). Therefore about 530 locomotive classes had become extinct since 1922. After 1947 no more than sixteen new steam locomotive classes were introduced on the nationalised British Railways before 1956.

That said, a significant number of locomotives that had been built during the Edwardian period, particularly heavy goods engines, lasted well into the 1960s. As far as they were concerned the end came on 8 September 1967, when BR Class Q6 0-8-0 No. 63344, built at Darlington Works as NER Class T2 No. 1251 in February 1913, finally ceased operation in Co. Durham after a working life of fifty-four-and-a-half years. Also there on the same day several former NER Class P3 engines (LNER J27) brought the 140-year 0-6-0 era to an end, which had also begun in Co. Durham when Timothy Hackworth had built the *Royal George* at Shildon for the Stockton & Darlington Railway in 1827. Just under a year later all *normal* steam locomotive working on the British standard gauge railway network ended for ever.

   * For the four grouped railways on their formation the total was 886, but the Met, MGNJR and SDJR, which at this point remained unaffected, collectively operated around twenty-five additional classes that later passed into LMSR and LNER stock between 1930 and 1937.

# Acknowledgements

The writer is indebted to Brian Stephenson for making available some very choice illustrations from his remarkable photo archive, and also to Laurence Waters of the Great Western Trust, to Dave Harris of the Midland Railway Study Centre at Derby, and to Richard Carmen, likewise for their photographic assistance. In addition, I would also like to thank the staff in Search Engine, the research facility at the National Railway Museum, York, for their unstinting help during my numerous visits there, and to my wife Christine for her technical assistance in preparing the manuscript.

Philip Atkins, Harrogate, September 2019

# Appendix 1
## Locomotive Mileage Characteristics of Leading Railways, 1913

| Railway company | Route mileage | Loco stock | Miles (million) Passenger | Goods | Shunting | Other | Total engine |
|---|---|---|---|---|---|---|---|
| GWR | 2,678 | 3,070 | 29.64 | 20.17 | 16.92 | 6.50 | 73.23 |
| LNWR | 1,802 | 3,084 | 31.17 | 19.80 | 17.49 | 7.57 | 76.33 |
| NER | 1,698 | 1,998 | 16.57 | 12.31 | 13.74 | 3.29 | 45.91 |
| MR | 1,519 | 3,019 | 23.38 | 26.61 | 14.20 | 6.81 | 71.00 |
| NBR | 1.251 | 1,058 | 9.49 | 9.88 | 7.30 | 2.90 | 29.57 |
| GER | 1,107 | 1,274 | 14.77 | 8.61 | 5.08 | 2.12 | 30.58 |
| CR | 896 | 997 | 10.56 | 7.68 | 7.12 | 2.52 | 27.88 |
| LSWR | 879 | 937 | 14.76 | 3.75 | 4.27 | 2.10 | 24.88 |
| GNR | 678 | 1,345 | 12.49 | 10.22 | 5.81 | 2.57 | 31.09 |
| GCR | 628 | 1,352 | 10.81 | 9.94 | 10.29 | 2.60 | 34.30 |
| SECR | 624 | 719 | 13.30 | 2.44 | 3.41 | 2.11 | 21.34 |
| LYR | 532 | 1,577 | 10.61 | 5.53 | 11.00 | 4.10 | 31.24 |
| GSWR | 447 | 521 | 4.88 | 3.25 | 2.52 | 1.56 | 12.21 |
| LBSCR | 430 | 600 | 10.26 | 1.94 | 1.94 | 1.10 | 15.24 |
| HR | 411 | 152 | 1.76 | 1.14 | 0.45 | 0.52 | 3.87 |
| GNSR | 335 | 117 | 1.54 | 0.71 | 0.40 | 0.19 | 2.84 |
| NSR | 206 | 188 | 1.58 | 1.36 | 1.41 | 0.21 | 4.56 |
| FR | 115 | 132 | 0.85 | 0.88 | 0.90 | 0.34 | 2.97 |
| TVR | 112 | 251 | 0.93 | 1.66 | 1.85 | 0.40 | 4.84 |
| HBR | 79 | 170 | 0.37 | 1.79 | 0.94 | 0.27 | 3.37 |

Additional electrically worked passenger mileage (million): NER 1.26, MR 0.98, LSWR 0.18, LYR 1.77 and LBSCR 1.42.

# Appendix 2
## Traffic Receipts v Locomotive Expenses, Leading Railways, 1913

(Abstracted from *Board of Trade Annual Railway Returns, 1913*)

| Rly | Gross traffic receipts, £m (R) Passenger | Goods | Total | Locomotive expenses, £m (L) Running | Repair | Total | L/R |
|---|---|---|---|---|---|---|---|
| GWR | 7.286 | 7.887 | 15.173 | 1.958 | 0.865 | 2.823 | 0.186 |
| LNWR | 7.214 | 8.865 | 16.079 | 2.321 | 0.728 | 3.049 | 0.190 |
| NER | 3.819 | 7.375 | 11.194 | 1.351 | 0.676 | 2.027 | 0.181 |
| MR | 4.900 | 9.107 | 12.787 | 1.924 | 0.713 | 2.637 | 0.206 |
| NBR | 1.962 | 3.240 | 5.202 | 0.767 | 0.217 | 0.984 | 0.189 |
| GER | 3.239 | 2.703 | 5.942 | 0.868 | 0.220 | 1.088 | 0.183 |
| CR | 2.094 | 2.975 | 5.069 | 0.765 | 0.206 | 0.991 | 0.196 |
| LSWR | 3.554 | 1.697 | 5.251 | 0.807 | 0.250 | 1.057 | 0.201 |
| GNR | 2.488 | 3.441 | 5.929 | 1.001 | 0.321 | 1.332 | 0.225 |
| GCR | 1.307 | 3.589 | 4.896 | 0.869 | 0.354 | 1.223 | 0.250 |
| SECR | 3.680 | 1.175 | 4.855 | 0.736 | 0.191 | 0.927 | 0.191 |
| LYR | 2.878 | 3.610 | 6.488 | 0.956 | 0.338 | 1.294 | 0.199 |
| GSWR | 0.902 | 1.079 | 1.981 | 0.317 | 0.135 | 0.452 | 0.228 |
| LBSCR | 2.636 | 0.860 | 3.496 | 0.524 | 0.168 | 0.692 | 0.198 |
| HR | 0.336 | 0.220 | 0.586 | 0.107 | 0.031 | 0.138 | 0.235 |
| GNSR | 0.272 | 0.245 | 0.517 | 0.082 | 0.019 | 0.101 | 0.195 |
| NSR | 0.293 | 0.723 | 1.016 | 0.121 | 0.058 | 0.179 | 0.176 |
| FR | 0.161 | 0.411 | 0.572 | 0.076 | 0.023 | 0.099 | 0.173 |
| TVR | 0.245 | 0.722 | 0.967 | 0.152 | 0.055 | 0.207 | 0.214 |
| HBR | 0.263 | 0.615 | 0.878 | 0.098 | 0.043 | 0.141 | 0.161 |

NB. In 1913 £1 was equivalent to £112 at 2018 price levels.

# Appendix 3
## Distribution of Ten Most Numerous Locomotive Wheel Arrangements on Britain's Leading Railways on 31 December 1913

| | 0-6-0 | 0-6-0T | 4-4-0 | 0-6-2T | 0-4-4T | 2-4-2T | 2-4-0 | 0-8-0 | 4-6-0 | 4-4-2T |
|---|---|---|---|---|---|---|---|---|---|---|
| GWR | 562 | 1111 | 368 | [26] | | 38 | 133 | | 123 | 31 |
| LNWR | 767 | 302 | 290 | 380 | | 269 | 170 | 376 | 340 | 50 |
| NER | 797 | 240 | 208 | 102 | 183 | 60 | 81 | 120 | 65 | |
| MR | 1495 | 388 | 393 | [14] | 231 | | 275 | | | [70] |
| NBR | 519 | 147 | 166 | 53 | 30 | | 16 | | | 30 |
| GER | 420 | 256 | 171 | | 58 | 239 | 115 | | 19 | |
| CR | 346 | 183 | 139 | | 131 | 1 | 35 | 8 | 36 | |
| LSWR | 115 | 55 | 302 | | 215 | | | | 26 | |
| GNR | 411 | 248 | 136 | 56 | 58 | | 68 | 55 | | 60 |
| GCR | 445 | 93 | 150 | 206 | [6] | 49 | 23 | 89 | 43 | 52 |
| SECR | 212 | 46 | 237 | | 215 | | | | | |
| LYR | 530 | 249 | 73 | 31 | 1 | 332 | 1 | 197 | 20 | |
| GSWR | 196 | 10 | 176 | | 21 | | 11 | | 19 | |
| LBSCR | 67 | 89 | 58 | 134 | 36 | | | | | 62 |
| HR | 12 | 6 | 77 | | 5 | | 4 | | 31 | |
| GNSR | | 9 | 97 | | 9 | | | | | |
| NSR | 57 | 51 | 5 | 28 | | 6 | 3 | | | 7 |
| FR | 57 | 19 | 18 | 21 | | 6 | 7 | | | |
| HBR | 85 | 31 | 5 | 18 | | | 10 | 15 | | |
| TVR | 75 | 14 | | 151 | | | | | | 6 |
| Other | 142 | c. 150 | 109 | c. 180 | 34 | 4 | 3 | 4 | | 4 |
| Total | 7310 | c. 3700 | 3178 | c. 1400 | 1233 | 1004 | 955 | 864 | 722 | 455 |
| % | 30.9 | 15.6 | 13.4 | 5.9 | 5.2 | 4.2 | 4.0 | 3.7 | 3.1 | 2.0 |

[ ] Recently absorbed from another railway company.

# Appendix 4
## *Select* List of Preserved British Locomotives Designed or Introduced During 1901–14 (locations as 2019)

SECR Class D 4-4-0 No. 737 (1901), National Railway Museum, York

MR Compound 4-4-0 No. 1000 (1902, rebuilt 1914), Barrow Hill Roundhouse

GWR City class 4-4-0 No. 3440 *City of Truro* (1903), STEAM, Swindon

NBR Glen class 4-4-0 No. 256 *Glen Douglas* (1913), Riverside Museum, Glasgow

GNR Class C1 4-4-2 No. 251 (1902), Locomotion, Shildon

GER, Class 0-6-0 No. 1217 (1905), Barrow Hill Roundhouse

MR Class 4 0-6-0 No. 3924 (1920) †, Keighley & Worth Valley Railway

(NER) Class P3 0-6-0 No. 2392 (1923), Keighley & Worth Valley Railway

GWR 53XX class 2-6-0 No. 5322 (1917), Didcot Railway Centre

(SECR) Class N 2-6-0 (Southern Rly No. 1874) (1925), Swanage Railway

GWR Star class 4-6-0 No. 4003 *Lode Star* (1907), National Railway Museum, York

NER Class T2 0-8-0 No. 2238 (1918), North Yorkshire Moors Railway

GWR 2-8-0 No. 2818 (1905), STEAM, Swindon

GCR Class 8K 2-8-0 No. 102 (1912), Great Central Railway, Loughborough

SDJR 2-8-0 No. 88 (1925) †, West Somerset Railway

SECR Class H 0-4-4T No. 263 (1905), Bluebell Railway

LTSR 4-4-2T No. 80 *Thundersley*, (1909), Bressingham Steam Museum

GER Class S56 0-6-0T No. 87 (1904), Bressingham Steam Museum

SECR Class P 0-6-0T No. 178 (1910) †, Bluebell Railway

NSR Class New L 0-6-2T No. 2 (1923), Foxfield Railway

(GER) Class L77 0-6-2T (LNER No. 999) (1924), East Anglian Railway Museum, Wakes Colne

GWR 42XX class 2-8-0T No. 4270 (1919) †, Gloucestershire & Warwickshire Railway

### Replicas:

In 2019 a working replica of a GWR Saint class 4-6-0, No. 2999 *Lady of Legend,* was completed at the Didcot Railway Centre.

At the time of writing a replica of an LBSCR Class H2 4-4-2 is under advanced construction by the Bluebell Railway. This is to be named *Beachy Head* and numbered (BR) 32424, and it will commemorate the last British 'Atlantic' to remain in service in this its final form.

† Other examples of this class have also been preserved elsewhere in the United Kingdom.

# Index